"As with Mike Hulme's career, this book ranges between the natural sciences, the social sciences, and the humanities. In doing so, it offers an interdisciplinary – and explicitly geographical – perspective on the 'key idea' of climate change. In 10 thoughtful chapters, Hulme opens up and extends understanding of the ways in which the idea of climate change is mediated through culture and politics. Selected key readings, provocative questions and scholar portraits increase the book's usability. I look forward to using it in my teaching practice."

Saffron O'Neill, *Associate Professor, Department of Geography, University of Exeter, UK*

"Is there a more contemporary 'key idea' than climate change? In this compelling overview, Hulme tracks how ideas of climate change have varied in space and time, and across cultural groups. From art to religion, from scepticism to cli-fi, he contextualises (and challenges) the matter-of-factness of a scientific view of climate change. Whether new to the topic or in need of a refresh, both students and senior scholars will find much of value here."

Lesley Head, *Redmond Barry Distinguished Professor, University of Melbourne, Australia*

"There is no one better qualified than Mike Hulme to explain the past, present and future of climate change in just ten chapters. In clear and engaging prose, Hulme leads us through the many facets of climate change: as a scientific concept, a locus of political debate, and a catalyst for imagined futures."

Rebecca Lave, *Professor and Chair, Department of Geography, Indiana University, USA*

"Mike Hulme's ground breaking writings have been the must-read texts on the social meaning of climate from theories of human difference, markers of place (those salubrious climates!) to science and technology studies. His work illuminates the conflicts, meanings, impacts and politics of climate change. By placing the understanding of climate as a socio-cultural as well as a scientific project, Dr. Hulme's work, always warm, generous and clearly written, has defined what it means to be an interdisciplinary, engaged scholar on a hyper-controversial topic. This magisterial book integrates climate questions through multiple discourses and controversies. Since it is hard to imagine a future without imagining climate change, this volume recasts and clarifies the nature of the debates. I think it is an essential volume for understanding atmospheric disorder, in all the meanings of the term."

Susanna B Hecht, *Professor, Luskin School of Public Affairs, and Institute of the Environment, University of California, Los Angeles, USA and Professor, International History, Graduate Institute for Development Studies, Geneva, Switzerland*

"This powerful and important book cogently demonstrates the need to take our ideas about climate change very seriously. Hulme shows the importance of recognizing climate change as a cultural predicament to be addressed through the explicitly performative mobilisation of different and competing values, ideologies, and narratives rather than a problem to be solved through more and better science and technology alone. An essential read."

John Robinson, *Professor, Munk School of Global Affairs and Public Policy, University of Toronto, Canada*

"This is a unique book with a truly interdisciplinary and comprehensive approach to the key ideas of climate change, and an all-in-one but concise reading of various ideas about climate change from social sciences, humanities and natural sciences. It is suitable for students and general readers trying to understand the profound climate changes. An innovative contribution of a human geographer to climate change studies!"

Weidong Liu, *Professor in Economic Geography, Chinese Academy of Sciences, Republic of China*

"What does climate change really mean for diverse communities? In this marvellous book, Mike Hulme explores the multitude of our human experiences of a changing climate. As a leading climate scientist, Hulme takes the reader beyond the science in a confronting, and profoundly enriching way. Building on a lifetime of climate research and the insights of marginalised voices, including indigenous, feminist, artistic, and religious insights, Hulme help us understand what it really means to be alive in a changing climate . . . to resist, struggle and imagine new futures, expanding our imagination in politically powerful ways."

Bronwyn Hayward, *Professor of Political Science, University of Canterbury, New Zealand*

"This book offers the most complete collection of key debates and examples from around the world that epitomizes the multifaceted nature of climate change. Reading it was for me an intellectually stimulating learning curve as Mike Hulme inspiringly reflects upon our personal and social bonds with the matter and idea of climate. Beautifully written, thought-provoking and easily accessible, *Climate Change* is the ultimate companion, and indeed a profoundly rewarding journey, for scholars of all disciplines."

Chaya Vaddhanaphuti, *Lecturer, Department of Geography, Chiang Mai University, Thailand*

CLIMATE CHANGE

Written by a leading geographer of climate, this book offers a unique guide to students and general readers alike for making sense of this profound, far-reaching, and contested idea. It presents climate change as an idea with a past, a present, and a future.

In ten carefully crafted chapters, *Climate Change* offers a synoptic and inter-disciplinary understanding of the idea of climate change from its varied historical and cultural origins; to its construction more recently through scientific endeavour; to the multiple ways in which political, social, and cultural movements in today's world seek to make sense of and act upon it; to the possible futures of climate, however it may be governed and imagined. The central claim of the book is that the full breadth and power of the idea of climate change can only be grasped from a vantage point that embraces the social sciences, humanities, and natural sciences. This vantage point is what the book offers, written from the perspective of a geographer whose career work on climate change has drawn across the full range of academic disciplines. The book highlights the work of leading geographers in relation to climate change; examples, illustrations, and case study boxes are drawn from different cultures around the world, and questions are posed for use in class discussions.

The book is written as a student text, suitable for disciplinary and inter-disciplinary undergraduate and graduate courses that embrace climate change from within social science and humanities disciplines. Science students studying climate change on inter-disciplinary programmes will also benefit from reading it, as too will the general reader looking for a fresh and distinctive account of climate change.

Mike Hulme is Professor of Human Geography in the Department of Geography at the University of Cambridge, and Fellow of Pembroke College, UK. The focus of his research career has been the analysis and explanation of the idea of climate change and his work has been published extensively across the natural sciences, social sciences, and humanities. He is the author of nine books on climate change, including *Contemporary Climate Change Debates: A Student Primer* (Routledge, 2020). From 2000 to 2007 he was Founding Director of the Tyndall Centre for Climate Change Research, based at the University of East Anglia.

KEY IDEAS IN GEOGRAPHY

Series editors: Noel Castree, University of Wollongong and Audrey Kobayashi, Queen's University

The *Key Ideas in Geography* series will provide strong, original and accessible texts on important spatial concepts for academics and students working in the fields of geography, sociology and anthropology, as well as the interdisciplinary fields of urban and rural studies, development and cultural studies. Each text will locate a key idea within its traditions of thought, provide grounds for understanding its various usages and meanings, and offer critical discussion of the contribution of relevant authors and thinkers.

MOBILITY, SECOND EDITION
Peter Adey

CITY, SECOND EDITION
Phil Hubbard

RESILIENCE
Kevin Grove

POSTCOLONIALISM
Tariq Jazeel

NON-REPRESENTATIONAL THEORY
Paul Simpson

CLIMATE CHANGE
Mike Hulme

For more information about this series, please visit: www.routledge.com/series/KIG

CLIMATE CHANGE

Mike Hulme

Routledge
Taylor & Francis Group
LONDON AND NEW YORK

First published 2022
by Routledge
2 Park Square, Milton Park, Abingdon, Oxon OX14 4RN

and by Routledge
605 Third Avenue, New York, NY 10158

Routledge is an imprint of the Taylor & Francis Group, an informa business

British Library Cataloguing-in-Publication Data
A catalogue record for this book is available from the British Library

Library of Congress Cataloging-in-Publication Data
A catalog record for this book has been requested

ISBN: 978-0-367-42202-8 (hbk)
ISBN: 978-0-367-42203-5 (pbk)
ISBN: 978-0-367-82267-5 (ebk)

Typeset in Joanna
by Apex CoVantage, LLC

TO EMMA JANE, WHO HAS LIVED WITH ME
"THROUGH" THIS BOOK.

CONTENTS

FIGURES

TABLES

BOXES

ACKNOWLEDGEMENTS

Parts of the subsection "Drama" in Chapter 7 (pp. 161–162) are drawn from Chapter 8 of Hulme (2016); some of the material in the subsection "Religious thought" in Chapter 8 (pp. 177–183) is based on Hulme (2017); and parts of the subsection "Who is governing" in Chapter 9 (pp. 208–214) are based on Chapter 11 of Hulme (2016). The Epilogue is an expanded version of a short essay first appearing in the *Scottish Geographical Journal* (2020: vol., 136, pp. 118–122).

I would like to thank Noel Castree, Editor-in-Chief of the Routledge *Key Ideas in Geography* Series, for his initial invitation to me. Without this prompt the book would not have been written. The observations and suggestions of five anonymous reviewers of the original proposal are acknowledged. They helped to sharpen the structure and balance of the book, as too did the suggestions of Noel Castree and one anonymous reviewer on an early draft of the whole book. I am also indebted to the following individuals for reading one or more chapters in draft and for offering valuable comments and suggestions that helped improve the text: Shin Asayama, Liz Astbury, Rob Bellamy, Kari de Pryck, Sarah Dry, David Durand-Delacre, Maximilian Hepach, Tayler Meredith, Sara Miglietti, Noam Obermeister, Anshu Ogra, Sarah Radcliffe, Ricardo Simmonds, Tom Simpson, Chaya Vaddhanaphuti, and Chao Xie. The usual disclaimer applies in that any errors or views expressed here are my responsibility alone. I also acknowledge the value of the regular STS reading group meetings in Pembroke College, where Shin, Noam, Maximilian, David, Kari, Freddie, Tom, Luke, and Christian offered a great sounding board to explore some of my ideas for presenting climate change. I could not have completed this book without the constant encouragement and support of my wife, Gill, not least through the weeks of Coronavirus lockdown.

I would also like to thank Ellie Fox who read through the entire manuscript from the perspective of a Gen Z student and offered many helpful suggestions to clarify my text and to avoid some egregious errors. Philip Stickler redrew in timely manner a number of the maps and graphs.

The following institutions and individuals are thanked for granting permission to reproduce assorted photographs, graphs, and images: The University of Kentucky (Figure 1.3), Tim Osborn (Figure 2.1, graph), Ulf Büntgen and Paul Krusic (Figure 2.2), Cian Scollard and the University of Vienna Library (Figure 2.3), Esteve Corbera (Figure 2.5), Roger Pielke Jr. (Figure 3.1), the *Financial Times* syndicate (Figure 3.2), Michael Sloan (Figure 4.2), Sheila Stickley and Jennifer Johnson (Figure 5.2), Tom Thornton and Glenn Rabena (Figure 6.1), Chris O'Brien, Kathy (Gotha) Guthadjaka and the Northern Institute at Charles Darwin University (Figure 6.2), Dehlia Hannah (the cover image and Figure 7.1), Diego Galafassi (Figure 7.2), Catharina Bonorden at the neugerriemschneider gallery (Figure 7.3), Harriet Hawkins (Figure 7.4), Mark Lawrence (Figure 9.1), Lisa Vanhala and Joana Setzer (Figure 9.4), and Reno Knutti (Figure 10.2). The poem *Höfn* by Seamus Heaney was reproduced with permission from Faber and Faber.

At Routledge, I thank Egle Zigaite, Andrew Mould, and also those involved in reading and production.

The front cover for this book is a cosmographic designed by Dehlia Hannah for her 2018 edited book *A Year Without a Winter*, published by Columbia University Press. It is used here with the author's kind permission and should not be reproduced. The image should be read counterclockwise, beginning with the cherub blowing down a cloud of air. Each figure signifies a calendar month.

GLOSSARY

The Anthropocene claims to define, in scientific terms, Earth's most recent geologic time period as one that is human-shaped, i.e., anthropogenic. This claim is based on overwhelming global evidence that atmospheric, geologic, hydrologic, biospheric, and other Earth System processes are now being altered by human technologies.

Black Swan events in public life are very rare, come as a surprise, but carry severe consequences. Their occurrence is often inappropriately rationalised with the benefit of hindsight. The theory was developed by Nassim Nicholas Taleb in 2007.

Carbon capture and storage describes the process of capturing waste carbon dioxide usually from large industrial point sources, such as a cement factory or biomass power plant, transporting it to a storage site and depositing it where it will not enter the atmosphere, normally an underground geological formation.

Climategate refers to a controversy that began in November 2009 with an unauthorised hacker accessing a server at the Climatic Research Unit at the University of East Anglia, UK. Thousands of emails (and computer files) containing correspondence between international climate scientists dating back more than ten years were uploaded to various internet locations and were scrutinised for evidence of scientific malpractice.

Cognitive dissonance occurs when a person holds contradictory beliefs, ideas, or values. It can also be experienced as psychological stress – dissonance – when they participate in an action that goes against one or more of their beliefs.

Critical realism distinguishes between the "real" world and the "observable" world. The "real", material world exists independently of human perceptions, theories, and constructions. But the world as we know and understand it can only be *constructed* from our perspectives and experiences, in other words through what is "observable" to us. And these observations are, to varying degrees, influenced by fallible social and technological systems and devices. We cannot get direct, unmediated access to reality.

Declension-ism/ist is a belief in the long-term decline of human culture and its associated capacities and impacts. It is a belief that may be applied to specific social or cultural formations, nation states, or, indeed, to the species as a whole.

Deliberative democracy is a form of democracy in which deliberation is central to decision making. It adopts elements of both consensus decision making and majority rule. Deliberative democracy differs from traditional democratic theory in that authentic deliberation, not mere voting, is the primary source of legitimacy for the law.

Eco-criticism is the study of literature and the environment from an interdisciplinary point of view, in which literature scholars analyse texts that illustrate environmental concerns and examine the various ways literature treats the subject of nature. **Material eco-criticism** emphasises the additional role of matter – for example, weather, rocks, planets, water – in giving meaning to the human experience of nature.

Ecosystem Services (see **Natural Capital**)

Ethics (of care, utilitarianism, and deontological). The **ethics of care** is a normative ethical theory, notably developed by feminist Carol Gilligan, that holds that moral action centres on interpersonal relationships and on care or benevolence as a virtue. Ethics of care emphasises the importance of response to the individual, in contrast to utilitarianism and deontological ethics that emphasise generalisable standards and impartiality. **Deontological ethics** holds that the morality of an action should be based on whether that action itself is right or wrong under a series of rules, rather than based on whether the consequences of the action maximize happiness and well-being for all affected individuals, which is **utilitarianism** (or consequentialist ethics).

Gaia (hypothesis) is the mythological name given to the Earth by James Lovelock, formally in 1979. It emerges from his argument that all living organisms and inorganic material are part of a dynamical system that shapes the Earth's biosphere and that maintains the Earth as a fit environment for life.

General purpose technologies are technologies that can affect an entire economy and that have the potential to dramatically alter societies through their impact on pre-existing economic and social structures. Examples would be the steam engine, railways, electricity, the automobile, the internet, CRISPR gene-editing, and artificial intelligence.

Global Warming Potential (GWP) defines the heat absorbed by any greenhouse gas in the atmosphere, expressed as a multiple of the heat that would be absorbed by the same mass of carbon dioxide. The GWP of carbon dioxide is 1. For other gases, their GWP depends on the time frame over which the calculation is made. Thus, methane has a GWP of between 30 and 80 and for nitrous oxide between 250 and 300.

Governmentality describes an approach to the study of power that emphasizes the governing of people's conduct through positive means rather than through disciplinarian forms of sovereign power. Governmentality is generally associated with the willing participation of the governed. It was initially proposed by French social philosopher Michel Foucault (1926–1984).

Heuristics are techniques – often mental shortcuts – that ease the cognitive burden of making a decision or solving a problem. They offer practical methods for reaching a quick decision or understanding of a problem but that may be suboptimal, imperfect, or biased. Applying "stereotypes" would be one example of a heuristic.

Infrastructural globalism is a concept introduced by historian of science Paul Edwards that describes the phenomenon by which "the world" as a whole is produced and maintained – as both an object of knowledge and a unified arena of human action – through global infrastructures, such as satellites, the internet, or computer models.

Integrated assessment models (IAMs) are models that simulate the main features of society and economy, together with the biosphere and atmosphere, within a single modelling framework. IAMs are used to inform policy making, usually in the context of climate change but also in other areas of human and social development such as health and agriculture.

International Geophysical Year (IGY) was a worldwide program of geophysical research that was conducted from July 1957 to December 1958. It was directed towards a systematic study of the Earth and its planetary environment and encompassed research in fields such as geomagnetism, glaciology, meteorology, oceanography, seismology, and solar activity.

Keynesian economics refers to an economic theory that advocates for increased government expenditures and lower taxes to stimulate demand and pull an economy out of depression. Named after British economist John Maynard Keynes (1883–1946), whose ideas were first influential in the 1930s and 1940s.

Lifeworld is a description of the world as immediately or directly experienced in the subjectivity of everyday life. It is usually distinguished from the objective "worlds" revealed by the sciences. People's lifeworld is shaped by individual, social, perceptual, and practical experiences. The idea was introduced by the German philosopher Edmund Husserl (1859–1938).

Likert scales are commonly used in research that uses survey questionnaires to elicit people's beliefs or perceptions. Likert scales require respondents to answer questions using a scoring range from one extreme attitude to another. There is normally a neutral option in the middle and so the scoring range is usually an odd number. It is named after the American social psychologist Rensis Likert (1903–1981).

Living Labs – or living laboratories – is a research-oriented concept, which may be defined as a user-centred, open-innovation ecosystem, often operating in a territorial context (for example, a city, a community). The Living Lab allows the co-creation, exploration, experimentation, and evaluation of innovative ideas, scenarios, concepts, and related technological artefacts in real life communities.

Material eco-criticism (see **Eco-criticism**)

Moral universalism holds that what is right or wrong is independent of local custom or opinion (hence opposed to **relativism**). Moral universalism is compatible with both moral absolutism (that considers all actions are intrinsically right or wrong) and also consequentialism (actions are judged according to their consequences).

Nationally Determined Contributions (NDCs) were established by the Paris Agreement on Climate Change and lie at the heart of achieving its long-term goals. NDCs embody efforts by each signatory country to the Agreement to reduce their emissions of greenhouse gases by an amount it has voluntarily determined and to adapt to the impacts of climate change. NDCs must be prepared, communicated, and maintained across successive reporting cycles.

Natural Capital describes the world's stock of natural resources, which includes geology, soils, air, water, and all living organisms. Some natural capital assets provide people with free goods and services, often called **Ecosystem services**. The term was first used in 1973 by E. F. Schumacher (1911–1977) in his book *Small is Beautiful*.

The Naturalistic fallacy is the argument that things that are natural are good (e.g. because breastfeeding is natural, it must be good), whereas things that are unnatural are unethical (e.g. because baby formula milk is unnatural, it must be bad). But in logical-moral terms this is fallacious. Many natural things are bad, and many unnatural things are good. The philosopher David Hume made a similar point when he observed that one cannot get an "ought" from an "is": For example, just because smoking is harmful to your health (fact) does not in and of itself mean that you ought not to smoke (moral injunction).

Net-zero (carbon) emissions refers to achieving net-zero carbon dioxide emissions, within a specific jurisdiction, organisation, or unit of territory, by balancing carbon emissions (from various sources) with carbon removal (often through carbon offsetting or **Carbon capture and storage**) or else simply eliminating carbon emissions altogether.

Relativism is the view that truth and falsity, right and wrong, are products of differing local cultural conventions and social frameworks of assessment. Relativists believe that such claims can only be justified relative to particular frameworks of assessment; there is no universal vantage point from which to adjudicate the truth or falsity, rightness or wrongness, of a statement.

Rogation days are days of prayer and fasting in Western Christianity, often observed with public processions and supplication to God seeking protection from natural calamities such as adverse weather.

Romanticism (the **Romantics**) was an artistic, literary, musical, and intellectual movement that originated in Europe towards the end of the eighteenth century. Romanticism was characterized by its emphasis on emotion and individualism, as well as a glorification of all the past and nature. It was partly a reaction to the Industrial Revolution and the scientific rationalization of nature, both central components of modernity.

Schumpeterian refers to an economic theory that states that growth is driven by innovation and governed by a process of creative destruction in which new production units replace old ones. Named after the Austrian economist Joseph Schumpeter (1883–1950), it became popular in the late twentieth century.

Scientific naturalism is the idea or belief that only natural laws and forces (as opposed to supernatural or spiritual ones) operate in the universe. Adherents of scientific naturalism assert that natural laws are the only rules that govern the structure and behaviour of the natural world and that the changing universe is at every stage a product of these laws. Everything that is can be explained through natural laws discoverable by science.

Time-banks are informal institutions that allow you to spend one hour of time helping someone out, for example by mowing their lawn or by doing their shopping. For every hour spent, you earn an hour's help from them in return in assisting in some appropriate way.

Tipping points describe thresholds (with respect to the climate system) that when exceeded can lead to large changes in the state of the system. Potential tipping points have been suggested in the physical climate system and in impacted ecosystems and sometimes in both.

Washington Consensus describes a set of ten economic policy prescriptions considered to constitute the "standard" reform package promoted during the 1980s for crisis-wracked developing countries by Washington-based institutions such as the International Monetary Fund, World Bank, and US Department of the Treasury. The term was first used in 1989. The prescriptions included the expansion of market forces within domestic economies.

Wicked problems are social or political problems that are difficult or impossible to solve because of incomplete, contradictory, and changing requirements that are often difficult to recognize or define.

PROLOGUE

Not another climate book!

One might well ask why another book on climate change. Surely beyond the tens of thousands that have already been published on the topic there is little more to be said. The literature on climate change is vast. It ranges from the mega-volumes that emerge every six or seven years from the United Nations' Intergovernmental Panel on Climate Change (IPCC) and that summarise scientific, technical, and academic knowledge on the topic, to the thousands of books on climate change that approach it sector-by-sector (whether birds, cities, finance, aviation, and so on), region-by-region (for example the Arctic, mountain environments or small islands), or through the lens of individual countries (for example Russia, the United States, or South Africa).

Then there are the guides to climate change developed for specific audiences – whether for "dummies", children, planners, or environmental lawyers – and the innumerable "short introductions" to climate change that different publishing houses are keen to print. There is another entire genre of climate book, which is the polemical. These books seek to persuade their readers that climate change is either a hoax, a tragedy, a crisis, or a cataclysmic ending. Or that climate change is a solvable problem – offering a manual about *how* to solve it – or that it is a problem *beyond* solving. Over the years, too, a growing number of novels, short stories, plays, and poems inspired by the idea of climate change have been published, a subject for creative writers nicely surveyed by Axel Goodbody and Adeline Johns-Putra in their book *Cli-Fi: A Companion* (2018). And these are just books appearing in English. Many of these climate change titles and themes are replicated in other languages.

What more possibly is there to add?

There is always more to say because climate change is not "over". There is "further beyond" – *Plus Ultra*, the epigraph engraved by Spanish grandees on the Pillars of Hercules at the Straits of Gibraltar at the turn of the sixteenth century. Climates around the world continue to change and will do so for the indefinite future. (The idea of a stable climate has always been a convenient fiction, even if widely believed at various times past.) This brute fact about climate is bringing forth new ecological impacts, technologies, policies, social movements, philosophies, poems and novels, political dramas, and dilemmas. These very real ecological, social, cultural, and political developments need studying, explaining, and writing about. Given the growing understanding of how climate is deeply connected with human socio-technical development, it is not surprising that there is always more to learn – and therefore always more to say – about climate change. Climate change has today become a synecdoche – it "stands-in" – for the status and prospects of people's changing material, social, and cultural worlds. And these worlds are always in the making.

But there is another sense too in which there is always "further beyond". The phenomenon called "climate change" is not one thing – but many things. Or rather, even if scientists study climate change as a singular physical phenomenon, then how people apprehend it – how people come to know climate change, speak about it, and respond to it – is certainly multiple. On the one hand, the idea of climate change describes an objective reality. The character of weather and associated physical environments – in sum, the climate – really *is* different from what it was 100 years ago. But it is also true that the *idea* of climate change relates to subjective perceptions of being alive in a changing world. Climate change is therefore *one* thing – a physical reality. But it is also *many* things – a multitude of subjective human experiences. For example, how climate change is understood and experienced in Malawi is very different from how it is understood and experienced in Mongolia, Mexico, or Malta. And within Malawi what climate change means for the Finance Minister responsible for public investment is very different from what it means for female farmers struggling to feed their families or for university-trained ecological scientists studying fish habitats in Lake Malawi.

All this is to say that the meaning of climate change is never fixed, nor can it ever be exhausted. A final universal meaning of climate change cannot be discovered, least of all by scientific inquiry. Just as physical climates – and the human and non-human environments they shape and interact with – will continue to change around the world, so too will the significance and meanings that people attach to these changes.

And this is where geography matters. Geography is a discipline of inquiry and discovery that seeks to reveal and understand the interactions between people and their environments and how these interactions vary around the world. The discipline of geography is therefore well-placed to expose and explain these different meanings of climate change and their significance for how people think and act today. And how those meanings may change in the future.

This now allows me to answer the question I posed earlier: Why another book on climate change? This book offers a perspective on the idea of climate change informed by the discipline of geography and shaped by the sensibility of a geographer. It is therefore *not* the same account of climate change that one might get from a climate historian, an Earth System scientist, a political scientist, or a development economist. And neither is it simply a shorter version – a précis – of the latest IPCC report, for example the IPCC's 6th Assessment Report due to be published in 2021/2022. For the reasons I outline earlier there can be no single-volume textbook on climate change that offers a complete explanation or a final account of what climate change is or of what it means.

Of course, this provisionality is true also for geography and for geographers. Whilst drawing heavily upon the traditions of geographical thought, scholarship, and practice, the account of climate change I offer here remains just one way of grappling with this confounding idea (see Box 0.1). A different geographer, one from Bolivia or one from Indonesia, say, would offer a different account. It is therefore important at the outset to disclose *my* approach to explaining the idea of climate change.

BOX 0.1 HOW I APPROACH THE STUDY OF CLIMATE CHANGE

Climate change is in many ways a *confounding idea*; that is, it is an idea that causes confusion by not being easily grasped by either human senses or intuitions. Or to use a complementary metaphor, climate change is a *kaleidoscopic idea* in that it seems to comprise a constantly changing pattern or sequence of elements, both material and symbolic. The characteristics of climate change cut across several of the dualisms that have inscribed themselves in many of the academic disciplines. Thus climate change is known both by scientists and by lay-experts – but who are each attuned to recognise different forms of presenting evidence; climate change unfolds in the physical world but also exists in the human imagination; climate change can be quantified objectively, but its risks are apprehended or experienced subjectively; climate change can be represented in numbers but also through stories; facts and values easily – and inevitably – become entangled in narrative accounts of climate change.

This kaleidoscopic nature of climate change is why I believe the eclecticism of geography as a discipline offers a good vantage point from which to try to make sense of it in a more holistic manner than perhaps other disciplines may achieve. Geographers are disciplined to appreciate the distinctive contributions to knowledge about the world that are made by the sciences, social sciences, and the humanities, even if they are not all equally versed in – or personally committed to – these different knowledge-ways. Nevertheless, the eclecticism valorised by geography has shaped how I have studied climate change through my career. And it helps explain the analytical approach adopted in this book. It is the benefits of this eclecticism that I submit as the reason why taking a geographical approach to understanding the idea of climate change is beneficial for readers beyond the discipline of geography.

To be clear about my own position: I recognise the physical realities of changing climatic systems and patterns of weather that are detectable and, within limits, predictable through scientific measurement and representation. At the same time, I recognise the socially constructed dimensions of this knowledge. In other words, I am a **critical realist**, not a naïve one. I recognise that different forms of knowledge, whether expert or lay, offer different readings and interpretations of physical environments and that these knowledges are influenced by cultural and political histories and geographies. I am an epistemic pluralist. I recognise that there are different moral truths held by people but would nevertheless want to submit such truth-claims to systematic inquiry that tests their coherence, credibility, and validity. I am a qualified **moral universalist**, not a **relativist**. I recognise the difference between facts and values yet also know that in many, if not most, cases these can be hard to disentangle. And I believe there are non-material dimensions of human consciousness and subjectivity that are beyond explanation by scientific method, as conventionally understood. I am not a **scientific naturalist**.

This set of epistemic commitments begins to explain how I approach the study of climate change and the account I give of the idea in this book. Scientific inquiry yields a range of knowledge about the changing physical world, but this knowledge does not dictate how any given individual or social collective should act. One cannot extract an "ought" from an "is", the **naturalistic fallacy**. There is a complementarity between logos, mythos, and ethos, between rationality, story, and character (see Table 4.2). Facts and meanings are different things, but they belong together: i.e., factual truth and meaningful truth are not necessarily exclusive truths, even if one arrives at them in different ways. A changing climate is objectively real and the causes of this change now largely human-driven. But there are many different ways of framing

(Continued)

and interpreting the risks and challenges that are compatible with this reality. Of necessity, these framings and interpretations rely upon other forms of reasoning beyond the narrowly scientific and that have varying degrees of social legitimacy and effectiveness. Much talk about "climate change" is inevitably talk about all sorts of things that do not emerge directly from the scientific study of the physical processes of a changing climate.

Understanding the book

My point of departure is that the book was commissioned as part of the *Key Ideas in Geography* series from Routledge. As stated by the publisher, this book series

> provides strong, original, and accessible texts on important spatial concepts for academics and students working in the fields of geography, sociology and anthropology. . . . Each text will locate a key idea within its traditions of thought, provide grounds for understanding its various usages and meanings, and offer critical discussion of the contribution of relevant authors and thinkers.

This book in the series is therefore written *by* a geographer and, as suggested previously, *for* geographers and allied disciplines. But the book is also written for scholars and students who study climate change from other vantage points but who may gain by seeing how a geographer writes the story of climate change.

The story I tell here is inevitably shaped by my own academic career in British geography, which stretches back to the early 1980s. During these 40 years or so, I have studied climate change using approaches to inquiry and scholarship drawn from the natural sciences, social sciences, and humanities. Such epistemic and methodological eclecticism is for me one of the hallmarks of the discipline of geography (Box 0.1). During this career I compiled and analysed many observed global and regional climate datasets. I founded and led a national climate change research centre and founded – and continue to edit – an interdisciplinary review journal on climate change. I participated for several years as a member of the IPCC, making a "significant contribution" to it receiving a share in the Nobel Peace Prize in 2007, advised numerous public and private organisations about climate change, and engaged openly in policy and public debates about climate change in many different countries.

In different ways, these professional experiences have all contributed to my thinking about the idea of climate change.

My approach to the study of climate change has been rooted in the specific insights and traditions of geographical thought and inquiry into which I was first inducted as a student of geography in the late 1970s. Nevertheless, I adopt a particular intellectual stance *within* the discipline of geography that guides and shapes the book. It is best described as one inspired by the sub-field of *geographies of knowledge*. A geographer of knowledge seeks to understand how knowledge about the world is made, contested, authorized, and used with particular regard to situated historical, cultural, and political processes. She wants to explain the spatially differentiated ways in which knowledge claims emerge in places, how they become institutionalised, and how they travel through social and cultural worlds. In other words, how knowledge is socially constructed. And he wants to reveal how these knowledge claims are then contested by different political actors and challenged by other cosmologies and knowledge systems.

This approach to understanding the idea of climate change is one that I have developed over the last 15 years or so. It has been inspired by critical, cultural, and historical geography and by disciplines cognate to geography – science and technology studies, environmental history, and history and philosophy of science. The central arguments of the book and the structure that I have chosen to present those arguments – explained in the rest of this Prologue – can be traced back to this set of intellectual commitments. My intention with *Climate Change* is to show how a geographer – in particular how a geographer of knowledge – might understand the idea of climate change.

What is my argument?

What then am I offering? At the heart of my argument is the assertion that the idea of climate change is multifarious. There is no single comprehensive account of climate change that can do full justice to the physical manifestations, political discourses, and imaginative power of the phenomenon. This then leads me to conclude that "climate change" is not a problem that can be solved, any more than the ideas of democracy or freedom are ones that can be "solved". The world's climates cannot simply be put back to some pre-disturbance condition. Climates cannot be restored or repaired in this way – although this does not stop some from trying, as we shall see. Nor can the *idea* of climate change be decisively dismantled or discarded. The idea can and will evolve; in the future it will carry different meanings. But as an idea, climate change will endure.

This irreversibility and intractability of climate change is what is captured by the notion of "**a wicked problem**". This notion of "wickedness" means that the problem itself is impossible to incontrovertibly define. Is the problem of climate change that the Earth's radiation balance is dangerously disturbed? *In which case the solution* might be to spray particles into the stratosphere to create a sun-screen. Is the problem of climate change that there is too much carbon dioxide accumulating in the Earth's atmosphere and oceans? *In which case the solution* might be to scrub carbon dioxide out of the atmosphere and sequester it in durable products or geological strata. Is the problem of climate change the entrenched self-interest of the fossil fuel and other extractive industries that wield too much influence over the levers of power? *In which case the solution* might be to divest and dismantle such industries in favour of less centralised and more enlightened ones. Is the problem of climate change that the environmental externalities of goods and services are not properly reflected in market prices? *In which case the solution* might be to establish a global price for carbon. Is the problem of climate change that the prevailing models of economic growth and social well-being remain too closely wedded to indicators of material consumption? *In which case the solution* might be to reimagine what is meant by a desirable and worthwhile human life and to inscribe such convictions into new policy metrics. Is the problem of climate change rooted in the colonial past and its legacies, which continue to subject peoples, species, and environments to coercive and extractive powers? *In which case the solution* might be reparations for the resulting ecological and cultural loss and damage caused by these colonising behaviours.

Of course, some would say that climate change is all these things and more. All possible solutions need exploring. And in the chapters that follow you will encounter all of these arguments and propositions. Yet to say that all of these solutions are needed simply because "time is running out" would be to misunderstand the nature of the predicament that climate change presents. Climate change is better thought of as a composite problem, a meta-problem made up of many smaller ones that are overlapping and interlocking in various ways. And solutions to any of these constituent problems mobilise different and competing values and ideologies. Solving any one part of this composite problem (i.e., solving a first-order problem) simply displaces or creates problems elsewhere (i.e., creating second- or third-order problems). To finally "solve" climate change would in fact be to exhaust the political imagination; it would have created a world beyond improving.

There are three underlying and interconnected claims that lie at the heart of this argument. First, given the mobilising power of the idea of climate change across multiple social worlds, it is important to *challenge simplistic and historically emaciated accounts* of what climate change is understood to be. The idea of climate change has a deep history with many different cultural roots and meanings. I use the analogy of democracy. As political scientists and historians well know, the idea of "democracy" does not define one – and only one – way for how a polity organises itself to make fair and acceptable decisions.

To think thus of democracy would be frustrating for the individual and ultimately dangerous for a society.

My second claim is this: Injecting into public life different and richer accounts of what climate change is and what climate change *means* is invigorating for contemporary politics. The account of climate change I offer in the book *challenges the dangerous hegemony of a naturalistic climate science*. Dissenting from hegemonic accounts of climate change opens up new ways of framing climate–society relations. This is good for thinking through difficult issues like climate change and for improving the quality of decision making. Dissent is good for society, as Jerome Kagan, Harvard Emeritus Professor, explains: "Every society needs a cohort of intellectuals to check the dominance of a single perspective when its ideological hand becomes too heavy" (Kagan, 2009: 266). And it is good for the individual. "Dissent . . . enables us to think more independently and to speak our own truth, and it stimulates our thinking", argues psychologist Charlan Nemeth. "We become more inquiring, more divergent in our thinking, and more creative" (Nemeth, 2018: 201).

My third contention is that the idea of climate is performative, it has effects in the real world. How one comes to know climate – and the account one gives of its changes – is *never politically neutral nor without effect* on the social ordering of today's and tomorrow's worlds. One cannot separate the merely descriptive from the intentionally normative. Any description of what climate change is carries with it latent but preferential modes of acting in the world in ways that substantively *change* the world.

One of the enduring features of climate change in public life over the last few decades is its contested nature. These contests have included arguments about the veracity of climate science and the integrity of its scientists, about the allocation of historic responsibility for climate change between and within nations, about the efficacy of policy instruments being used to arrest climate change, about the wilful spread of disinformation about climate change and, more recently, about the language that should be used to talk about climate change in public. Some of these arguments are explored in a previous edited book *Contemporary Debates in Climate Change: A Student Primer* (Hulme, 2020), where I enlisted pairs of scholars from around the world to debate some of these issues.

But my purpose in writing *this* book is different. It is to establish the cultural origins of the idea of climate change and to help readers recognise some of the different accounts in public circulation around the world about what climate change means today. And it is to help readers comprehend how each account is rooted in different histories, ideologies, epistemologies, imaginaries, values, and hopes. Accomplishing this task of recognition is the guiding ambition of the book.

How is the book structured?

The book is structured in three sections, which broadly map onto the conventional Western temporality of past, present, and future, an approach first used with climate in Hubert Lamb's two-volume series *Climate: Present, Past and Future* (Lamb, 1972, 1977). Section 1 contains two historical-geographical perspectives on the idea of climate change. Chapter 1: *Climate and culture through history* develops the historical argument that climates and cultures have always been intimately related in the human imagination. It shows how this relationship has been understood in the past and how, over the years, it has been written about. In other words, Chapter 1 offers a brief intellectual history of the idea of climate change. Then in Chapter 2: *Climate change and science* I consider how scientific knowledge of the changing climate was developed through the political patronage of empires, superpowers, and, latterly, the United Nations. I take a geographies of science approach to interrogate the scientific kinds of climate knowledge that have emerged over the last 200 years. This knowledge has been constructed from the infrastructures, networks, and practices of science enabled through particular configurations of political power and sovereignty.

Section 2 of the book is concerned with the idea of climate change in "the present", by which I mean the last 30 years. During this period, the idea of climate change has circulated around the world and increasingly become a new locus for geopolitics. Each of the six chapters in this section advances a distinct way of public reasoning and expression through which climate change becomes a matter of public concern rather than simply a matter of scientific fact. I simplify some of the complexities involved in these moves by contrasting three "science-based" positions on the significance of climate change (Chapters 3, 4 and 5) with three "more-than-science" approaches for establishing its meaning (Chapters 6, 7 and 8). This is a somewhat crude distinction to make and it is explained and justified in the section introduction. But it is a helpful device for exposing how the idea of climate change becomes imbued with multiple meanings across diverse social formations, each drawing upon different epistemic and cultural resources.

The three respective positions on climate change outlined in Chapter 3: *Reformed modernism*, Chapter 4: *Sceptical contrarianism* and Chapter 5: *Transformative radicalism* are differentiated according to the convictions that people hold about two central considerations: The reliability and credibility of climate science and the desirability of the world's dominant capitalist political economy. The three positions of *reformism*, *contrarianism*, and *radicalism* are grouped together because in each case science is granted distinctive cognitive and cultural authority for knowing and acting in the world. It is not that they offer "science-only" positions. They each in different ways bring extra-scientific reasoning to bear on the question "what does climate change signify?" But they have in common a shared deference to science – at least a shared *rhetorical* deference – with regard to the search for an authoritative basis upon which to build their normative claims.

The following three chapters – Chapter 6: *Subaltern voices*; Chapter 7: *Artistic creativities*; Chapter 8: *Religious engagements* – explore what the idea of climate change means to social formations who adopt what I call "more-than-science" approaches in their search for meaning. These meanings are differentiated from those outlined in the preceding three chapters because they are grounded in the conviction that scientific knowledge provides an inadequate or insufficient basis for being, knowing and acting in the world. The claims of Western science about the changing climate are challenged, not on the grounds of their being necessarily "wrong". They are challenged on the basis that science provides neither the motivational energy nor the moral orientation needed to change the world.

Section 3 is concerned with climate change to come, how attempts to steer global climate are being made, and how future climates are being imagined. In Chapter 9: *Governing climate* I summarise the widening diversity of approaches used to "govern" future climate, for example through international treaties, carbon markets, the courts, technologies, citizens' assemblies, and social practices. These institutions are mobilised, at different scales, to serve the overarching ambition to alter the future trajectory of global climate. Finally, in Chapter 10: *Climate imaginaries* I consider how the idea of climate change creates both dystopic and utopic future worlds in the imagination. It introduces the idea of climate imaginaries and shows how these imaginaries are made salient through different futuring practices: Scenario planning and scientific modelling; metaphors and creative fiction. Finally, in a short Epilogue, I consider the future of the *idea* of climate change and relate these thoughts back to the discipline of geography.

Who should read this book?

Climate Change is written as an advanced textbook about the idea of climate change. It is suitable for undergraduate and graduate courses within the social sciences, humanities, and natural sciences that embrace climate change from interdisciplinary perspectives. As explained earlier, the book is commissioned in Routledge's *Key Ideas in Geography* series and so it clearly appeals directly to students in geography and directly allied subjects, such as environmental studies, environmental science, sustainability studies, anthropology, and development studies.

But the book is perhaps even more valuable for those studying climate change from across a much wider range of disciplines in the social sciences, humanities, and natural sciences – for example critical theory, earth sciences, ecology, engineering, history, international relations, linguistics, literary studies, media and cultural studies, philosophy, political science, religious studies, science studies, sociology, and many more. Science students studying climate change on interdisciplinary programmes will particularly benefit from reading the book to understand how the idea of climate change is understood from different epistemic traditions to those of natural science. And the relevance of *Climate Change* extends well beyond those students who happen to be studying in the UK higher academy. By drawing upon examples from many regions and across different cultures it offers a global geography of the idea of climate change, with which students from different world regions can engage.

Climate Change will also prove valuable to a wider readership than those formally studying for a higher education degree. For example, professionals, policy advocates, public communicators, teachers, and concerned citizens will find here a distinctive synoptic account of the idea of climate change. The book offers an orienting framework of understanding within which readers can locate their own position or that of public discourses with which they are familiar. I do not expect everyone to agree with my characterisation of climate change. Far from it. The account I offer cannot fully escape my own training, experiences, positionality, and values. Box 0.1 outlines my epistemic commitments; here, I mention my personal and cultural characteristics: A white, English, Christian, heteronormative male; and an economically radical, socially conservative, politically pragmatic, and intellectually curious professor of geography. At times my own biases will come through. Nevertheless, I seek to expose the multiple meanings that people in the world – and with different life histories and normative commitments to mine – attach to the idea of climate change.

Climate Change has a number of features to aid learning. Each chapter has a clear ending summary and a guide to selected further key readings through which to pursue some of the ideas raised. Each chapter also offers a few questions that may be suitable for classroom discussion, coursework essays, or as the inspiration for exam questions. Boxed explanations of key concepts, books, and institutions are provided, as well as short portraits of a number of scholars – especially geographers – who have made distinctive contributions to the understanding of climate change. A glossary of key terms and concepts is also included, together with a comprehensive bibliography and a detailed index.

Why now?

There is a shorter and a longer answer to this question. The short answer is that in 2018 I was invited to contribute this title to the *Key Ideas in Geography* series. There is no doubt that "climate change" is a key idea for geographers to grapple with and so the invitation immediately appealed to me.

The longer answer would be as follows. This book can be understood as the third contribution to a trilogy of books I have written about the idea of climate change. *Why We Disagree About Climate Change* (Hulme, 2009) was written in the late 2000s. It was an attempt to situate disputes about climate change – at the time heavily framed around the veracity of climate science – within a much broader set of considerations about how people orient themselves in the world: How people thought about science, economics, religion, risk, communication, development, and governance. The main ambition of that book was to decentre science from the story of climate change. Then, in the mid-2010s, I wrote *Weathered: Cultures of Climate* (Hulme, 2016), what might be considered a prequel to *Why We Disagree*. *Weathered* provided a deeper historical and cultural account of origins of the idea of climate and how the idea of climate continues to fulfil numerous political and cultural functions in today's world. Understanding climate comes before understanding climate change and so, logically, *Weathered* should be read before *Why We Disagree*.

Now, several years later, this book – *Climate Change* – can be seen as completing this trilogy. In one sense, nearly 15 years later, it updates the core argument of *Why We Disagree*. It has a similar ambition to this earlier book, namely explaining why climate change becomes a subject of intense political contestation. But it has a more carefully crafted focus on the *idea* of climate change, opening up the different ways this idea is interpreted by political movements and through different cultural practices. The book also highlights more decisively why climate change is not best understood as an engineering problem or even solely as a new locus for politics. Rather, it argues that climate change presents a predicament, a difficult, perplexing or trying situation for humanity. As with my other books, *Climate Change* is reflective rather than exhortatory. You will not find here simple solutions to climate change. I do not believe there are any.

Interestingly, each of these three books was written against the backdrop of major geopolitical events that have shaped – and continue to shape – the story of climate change in significant ways. *Why We Disagree* was written during 2007/2008, just as the world entered into its most serious financial crisis for more than 75 years, whilst at the same time preparing for a decisive international climate agreement at Copenhagen in 2009. *Weathered* was written during 2015/2016, immediately before and after the negotiation of the Paris Agreement on Climate Change. This agreement was described at the time by some as a "historic breakthrough" but by others as a "dismal failure".

And now this book, *Climate Change*, was written during 2020 as the world's most serious pandemic in a century unfolded, the full implications of which for the idea of climate change are far from being realised. The story of climate change therefore continues to be written – by political and social events, by the unfolding of physical processes on the planet and by the intervention of writers, scholars, poets, and playwrights. *Plus Ultra* – there is always "further beyond".

Mike Hulme
Pembroke College, Cambridge
December 2020

Section 1

CLIMATE HISTORIES, GEOGRAPHIES, AND KNOWLEDGES

One of the best-known works of English poet Philip Larkin is *Annus Mirabilis*. Written in 1967, Larkin picks on 1963 as the year to mark rhetorically when sex was invented in Britain. . .

> *Sexual intercourse began*
> *In nineteen sixty-three*
> *(which was rather late for me)-*
> *Between the end of the "Chatterley" ban*
> *And the Beatles' first LP.*

We might well ponder a similar question. When did climate change first "begin"? A standard scientific answer to this question might suggest the nineteenth century, when the effects of the fossil fuel energy revolution were first imposing themselves on the Earth's atmosphere. A different science-based answer might be that climate has always changed throughout Earth's history and so the question makes no sense.

So, let me rephrase the question. When did the *idea* of climate change first take hold of the human imagination? Did climate change only begin in any meaningful political sense in the 2000s with the emergence of transnational social movements campaigning to stop climate change? Or did climate change begin in 1988 with the formation of the United Nations' Intergovernmental Panel on Climate Change, the IPCC? Was its origin earlier than this, perhaps in the mid-1960s when the United States' presidential scientific advisory committee first warned President Lyndon B. Johnson of the dangers of human interference with the world's climate? Or maybe one might argue that the idea of climate change has always been with us. Some awareness of climatic instability has perhaps always been part of people's experience of their atmospheric surroundings, albeit an awareness expressed in very different ways at different times and places.

Many of the younger readers of this book may never have known a time when the idea of climate change did not seem to dominate world politics or to connect all parts of the planet. Certainly, the vibrant movement amongst school-age teenagers catalysed by Greta Thunberg during 2018 and 2019 engaged the first generation – Gen Z – that has grown up with the unavoidability of climate change. This generation has lived with climate change as a subject of study at school, as an everyday reality through social media, and as a matter of concern around which their anxieties about the future coalesce. Older readers who lived through the 1980s and 1990s may have first encountered climate change framed as an emergent environmental problem of modernity. Non-Western readers may place climate change in a different frame of reference, perhaps as an obstacle to development, as a legacy of colonialism, or as a reason – and a means – for restructuring relations between nations.

However you, the reader, may have encountered the idea of climate change – and within whatever frame of reference – it is important to place your current understanding of climate change, and your concerns about it, within a longer historical context. "Before changing the future, it is helpful to survey the past". Such a survey is the purpose of the two chapters in this section. I approach the question of climate change's *past* primarily through the disciplinary lenses of *historical geography* and *geographies of science*. Physical climates and their propensity to change have been known and interpreted in many different ways in past human cultures. Gen Z protestors are not in fact the first generation to encounter the idea of climate change, even if they may be the first to do so as an overtly political issue operating across borders in a globally connected world.

Some of these past interactions between climate and human culture are explored in Chapter 1: *Climate and Culture Through History*. This chapter illustrates the ways in which the idea of climate has been given shape and meaning in different human cultures. It argues that the idea of climate helps stabilise relationships between people's culture and their weather. The focus of Chapter 2: *Climate Change and Science* is the development over the past two centuries of an explicitly *scientific* form of knowledge about climates that change. The chapter outlines the changing political and cultural conditions that have enabled climate science to secure its dominant position in contemporary accounts of a changing global climate.

1

CLIMATE AND CULTURE THROUGH HISTORY

. . . Climate change historicised . . .

Introduction

This book is about the idea of climate change. It is a powerful idea that is re-making the contemporary world and shaping the twenty-first century. But climate change is not a new idea. It has existed in many guises in earlier societies and in different cultures. It is therefore important to start this investigation into the idea of climate change by suspending for a moment one's own beliefs about the phenomenon and to recognise different understandings. This is what this chapter seeks to do.

It is also necessary to recognise the cultural histories and geographies of the idea of *climate* itself, as distinct from climate change. Climate too is an enduring idea of the human imagination – but also a malleable one. Imagining climate as a singular global physical entity – to be studied and predicted through scientific means – is far from the only way in which it may be apprehended. Climate is an idea that people have used variously in different times and places to describe their experience of a well-ordered cosmos, to cultivate a sense of stability and social identity, and to provide explanations or justifications for cultural norms. For example, the idea of climate has been used to explain the "superior" culture of ancient Greece compared to Egypt, to regulate architectural designs, to justify slavery in America, and to guide military strategy – and much more besides. These cultural functions of the idea of climate emerge from people's place-based experiences of the weather and their search for order and meaning in an often turbulent and bewildering world.

It may seem strange for a book inspired by the discipline of geography to begin by talking about history. But understanding the past is a good starting point if one wishes to see how and why the idea of climate change today provokes such different reactions. Removing ourselves for a while from the present – and distancing ourselves from our own culture – may help broaden our grasp of what climate change might signify for different people today. And it might help us understand why the idea of climate change has become so powerful in today's world.

Since this chapter is about geographies and histories, it draws on the work of *historical* and *cultural geographers* and *climate* and *environmental historians*. I start by emphasising how climate, historically, has been an idea inescapably bound up with people's sense of place and dwelling, a central aspect of the geographical imagination. Just as physical climates vary spatially, so too does knowledge *about* climate. The physical climates of Bhutan and Bulgaria are very different. But how Bhutanese know their climate is also very different from how Bulgarians known *their* climate. The chapter then illustrates some of the different cultural functions of climate, how it comes to reflect cosmic order, how it acts as a descriptive index of aggregated weather, and how it has offered an explanation for human difference. I also consider the problem societies face when climates are perceived to change. What constitutes a culturally credible explanation for such atmospheric disorder? What duties, responsibilities, and capabilities do people have to restore climatic order? The chapter concludes with a brief historiography of climate change, how the idea of climate change has been written *about*. This highlights some of the ways that climate change has been written into histories of the past.

The idea of climate

Imagine you arrive in a foreign land with which you are totally unfamiliar. You have no prior knowledge of either its climate or its weather. Consider for example Māori of the fourteenth century arriving in voyaging canoes at Aotearoa New Zealand from their Polynesian islands. Or consider early English settlers arriving in sailing ships on the eastern seaboard of America in the early 1600s at a place they called Jamestown, Virginia. If you were a member of either of these migrating groups you would immediately encounter the weather of these unfamiliar places. But the weather would be strange to you; you would be uncertain about its behaviour tomorrow or next month. Your only register of atmospheric normality would be that inherited from your homeland, thousands of miles away. It would take you several years of settled life before you began to develop any sense of the seasonal rhythms of weather in these new places – what range of weather to expect, what extremes to regard as unusual, what patterns of change to expect from day to day, from season to season, or from year to year. Your sense of the *climate* of these places would develop only slowly, through lived experience. Only slowly would you become, one might say, acclimatised.

People always encounter their weather in particular places (Endfield, 2019). But as well as experiencing weather through their physical senses, people interpret weather through their imagination. This imposes a degree of order on their experience of the weather. And together, people construct shared collective knowledge – cultural knowledge – of how and why the weather of a particular place behaves the way it does. For this reason, people's experiences and imaginations of their weather are always infused with cultural meaning. For most human cultures, the idea of climate – and its many linguistic variants, such as *gihu* (Korea), *sila* (Inuit) or *fūdo* (Japan) – is simply the result of this search for order amidst the human experience of always changing and sometimes chaotic weather. This emergent sense of a place's climate – this attempt by human cultures to apprehend and describe the patterns of their weather – is universal. It is also ancient. Thus, in the Hindu Vedas one finds discussions of weather cycles, in the Hebrew scriptures the search for regularity in the weather is bound up with people's belief in a providential God, and in ancient Chinese court records one finds descriptions of patterns in the timing and frequency of extreme meteorological events. The idea of "climate" offers people a sense of regularity and stability of their atmospheric surroundings. Without this sense of climate to anchor their expectations of the weather, people would flounder in too bewildering an experience of atmospheric uncertainty and insecurity.

To complete this task of bringing order to weather, people draw upon many other aspects of their daily lives beyond their visceral experience of the elements of rain, wind, cold, and heat. People's ordering of their weather is therefore intimately bound up with manifestations of human cultural life. Their encounter with landscapes, their social and animal relationships, their livelihood practices, their imaginations, memories and stories, their gods and spirits . . . all these matter for the task of making sense of the weather. These cultural forms of life – this "ensemble of customary behaviours, institutions and artefacts" (Ingold, 1994: 329) – become resources in the project of weather ordering, of climate making. People's sense of climate therefore reveals as much about their culture as it does about their sensory experience of the atmosphere's weather. For the Inuit of West Greenland, "weather and climate are not variables out there to be measured and averaged, but result from an ongoing rhythmic exchange across substances – between earth and sky, organisms and their environment" (Simonetti, 2019: 259).

For this reason, the idea of climate is powerfully attached to people's sense of place and dwelling. Establishing the climate of one's locality is therefore hard-won. Climate only takes form and meaning through long-lived experiences of weather in particular places, experiences that are as much inflected by cultural norms and collective memories as they are by the specific life history of any individual. Affirming one's local climate becomes a mark of identity and belonging, as well as a means of differentiating between regions, countries, ethnicities, and livelihoods. For example, I live in the city of Norwich, which is in the East Anglian region of England. When asked about "my climate" I could legitimately describe the climate either of Norwich, of East Anglia or of England. But in each case, I would not only be saying something about where I lived but also different claims about who I was. Climate, then, can be understood as an idea comprising a constellation of weather sequences, physical environments, cultural memories, social identities, and practices.

This is shown nicely in one study of how international students living in Melbourne, Australia, struggled to readjust their everyday social lives and practices to the unfamiliarity of a novel climate (Strengers & Maller, 2017). And this is why many people, when asked to explain their climate, offer an account that greatly exceeds a narrow description of expected weather. For example, the Japanese word *fūdo* does not merely designate a climatic zone or region, as the English word "climate" might do when speaking, for example, of the climate of the Arctic. Rather "*fūdo* indicates an area of nature implicitly considered from a certain stand point, namely, insofar as it shapes or determines a culture" (Johnson, 2019: 18). Similarly, for Australian Aboriginals and Torres Strait Islanders, climate is connoted as "Country", an Aboriginal English word that embraces the weather, seasons, places, people, and their wider relationships, which, together, constitute a homeland.

The modernist notion of climate – invented and successfully mobilised by Western scientific culture over the last 200 years (see Chapter 2) – violates these richer cultural understandings of the idea of climate. It detaches explanations of the physical processes of the atmosphere from the cultural meanings ascribed by people in places to the patterns and regularities of their lived-in weather. One example of such violation is the use of "climatic spatial analogues". Spatial analogues have occasionally been used by scientists to communicate to public audiences in simple terms the extent of future climate change.[1] To translate scientific predictions of future climate change into supposedly meaningful human terms, today's climate of one place is used "to represent" the future climate of another place. Thus, for example, the present-day climate of Catalonia in northeast Spain might be used to represent the future climate of eastern England, or the future climate of Virginia will become like that of Florida today. Certain statistical properties of the future weather of a place may indeed come to resemble those of these analogue regions. But the wider set of place-bound cultural and historical attachments that contribute to a person's sense of climate are entirely lost through this use of spatial analogues. Climate and culture cannot be so easily detached.

The cultural functions of climate

The idea of climate, once formed, is bound up with how people live, think, and act. It is an idea that emerges as much from culture, as culture is something which emerges from climate. Unsurprisingly therefore, the idea of climate is called on to perform numerous cultural functions. Climate can act as a guide to agricultural and forestry practices; it can offer scope for thinking creatively about building design; it can act as a proxy for human health; it can provide apparently simple (naturalistic) explanations for complex (social) phenomena; it can be invoked to describe the essential character of a place. For these – and similar – reasons, the idea of climate is inescapably political. I have explored many of these cultural functions of climate in a previous book, *Weathered: Cultures of Climate* (Hulme, 2016). This subsection examines just three of them in some detail: Climate as a reflection of cosmic order; climate as a descriptive index of aggregated weather; and climate as a causal explanation.

Climate as cosmic order

In an essay that explores the boundaries of nature, historian of science Lorraine Daston explains that "without well-founded expectations, the world of causes and promises falls apart" (Daston, 2010: 32). As seen earlier, the idea of climate offers just such "well-founded expectations" of how the atmosphere will behave. Climate offers a way of navigating between the human experience of a constantly changing atmosphere – with its attendant insecurities – and the need to live with a promise of environmental stability and regularity. This disposition is what Nico Stehr (1997) refers to as "trust in climate". For people living in a place, climate offers an ordered container by which the unsettling arbitrariness of the restless weather is held in check. Climate can be "trusted" to define the boundaries of what weather might reasonably be expected to occur in a given place. This container creates Daston's required orderliness, which then stabilises cultural relationships between people and their weather.

But the idea of climate fulfils a deeper function than this. Climate is not just about the human ordering of the weather. Yes, the idea of climate tames the unpredictability of the atmosphere. But for many cultures the idea of climate is inescapably bound up with a wider set of well-ordered human relationships: With human kin, animals, spirits, and deities. As anthropologist Peter Rudiak-Gould has shown, when the Marshallese of western Oceania talk about the condition of their *mejatoto* ("climate"), they are in fact talking about the cosmic, meteorological, geological, oceanic, temporal, moral, cultural, economic conditions and relationships of their social lives (Rudiak-Gould, 2012). Changes in any of these conditions – whether the cost of living, the violation of kinship taboos, or offending the spirits – are sufficient to unsettle the climate. As their climate changes, so does their world. Changes in climate are experienced similarly among Indigenous communities across the Arctic. Things are experienced as acting strangely, as being out of sync, as when species and ice respectively arrive and retreat earlier than usual.

So, on the one hand, the belief in a settled climate enables the possibility of a stable psychological life and of meaningful human action in the world. On the other, a well-ordered social world or cosmos manifests itself in a stable climate. As explained by Thomas Ford, "'climate' is a term for a totality or encompassing environment of meaning" (Ford, 2016: 159). The idea of climate not only makes the atmosphere legible and predictable. It also holds together people's imaginative worlds. The nineteenth-century German naturalist Alexander von Humboldt's famous definition of climate as a description of the totality of ecological relationships of a place therefore did not go far enough. In the historical-cultural perspective offered here, the idea of climate reflects human belief in a well-ordered cosmos, refracted through the experience of place.

This particular function of the idea of climate – as a description of the totalising social experience of human life – seems far-removed from the scientific notion of climate that has emerged from Western rationality (see Chapter 2). And yet, as will be shown in later chapters, this world-ordering function of climate continues to resonate today for many people. It perhaps offers a better way of understanding many contemporary public responses to climate change than does a purely scientific account of the phenomenon.

Climate as index

Climate is widely understood today as "average" weather, a statistical description of an ensemble of weather experienced over a defined period of time in a specific place. Thus, the climate of my home city of Norwich, England, has an average annual temperature of about 10.2°C and an average annual precipitation of about 630mm. Climate understood thus – as a statistical descriptive index – is only possible because of systematic measurements of the physical properties of the atmosphere that first became common in eighteenth- and nineteenth-century Europe. Such a quantitative index establishes climate as a statistical entity, abstracted from the "lived" human experience of weather. Nevertheless, this indexical formulation of climate performs important cultural functions. Not least is its communicative value. Using formal comparative statistics, climates in different regions – and in different time periods – can be compared numerically rather than impressionistically. Thus, I can compare the climate of Norwich, Connecticut, USA, with my own city of Norwich, England. They have a very similar average temperature (10.1°C cf. 10.2°C), but Norwich, USA, is much wetter (1180mm cf. 630mm).

The benefits of this new statistical climatology emerged during the nineteenth century. For example, climate in this indexical sense began to be used as a valuable **heuristic** for resource planning by government agencies and for enabling new commercial projects. This use of climate aided European white settlers in unfamiliar lands of North America and new state bureaucracies such as the US Department of Agriculture in their task of planning

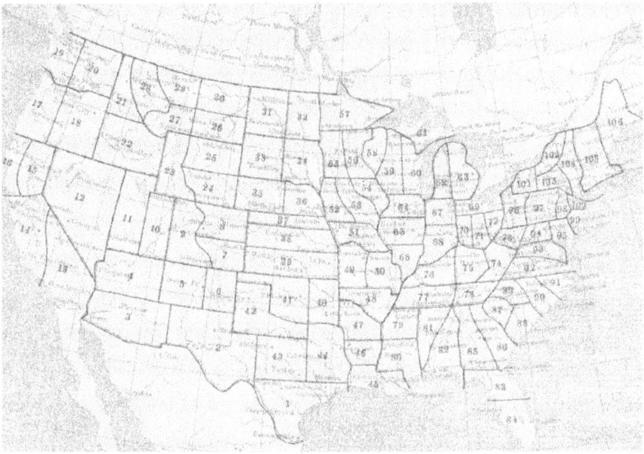

Figure 1.1 106 climatological sections in the United States as defined by US Department of Agriculture in 1911.
(*Source*: *Monthly Weather Review*, 1911).

and operating new agricultural systems (see Figure 1.1). This statistical definition of climate was enrolled in such projects because it offered a vision of numerically stable and (within limits) predictable weather that served the new demands of agricultural markets and financial investors (Baker, 2021). Order could be imposed on otherwise unreadable atmospheres and landscapes. Commercial investments in the newly settled lands of the American Midwest could be guaranteed a return. Similarly, the new insurance companies of the nineteenth century – whether hail insurance for farmers or life and travel insurance for European colonists – started using indexical climate as a basis for setting premiums and actuarial calculations (Kneale & Randalls, 2014). Statistical climates served the important commercial task of pricing risk. As James Scott explains in *Seeing Like A State: How Certain Schemes to Improve the Human Condition Have Failed* (1998), from the mid-nineteenth century onwards statistical climatology bolstered modernist ambitions of the state – often the imperial state – by making environments legible and people governable.

Climate as explanation

As well as reflecting a wider cosmic order or offering a descriptive index of the state of the atmosphere, the idea of climate has also served the human need for explanation. Climate has long been elevated as a chief explanation for a bewildering variety of environmental, social, individual, and political phenomena. Indeed, Fleming and Jankovic (2011: 2) in their account of the idea of climate suggest that "one is more likely to encounter climate as an agency [as explanation] rather than an index [as description]. Climate has more often been defined [by] what it *does* rather than [by] what it *is*" [emphasis added].

The story of climatic determinism – sometimes referred to as "climatic theory" – and its controversies is well known (Keighren, 2010). Academic geographers have had to wrestle with the difficult relationship between climatic and human agency ever since the discipline formed in the late nineteenth century (see Box 1.1 and the work of Ellen Semple). For philosophers of classical antiquity such as Herodotus and Hippocrates, the idea of climate had offered a convenient way of explaining differences in the physiology and psychology of individuals, cities, and peoples. These explanations appeared to offer naturalistic reasons for dividing up the human species into distinct races, languages, and cultures (Glacken, 1967). Climate was a useful device for differentiation. It offered alluring explanations for why forms of social organization and expressions of human behaviour vary from place to place. Thus tropical climates were claimed to cause laziness and promiscuity, while the bracing climates of the middle latitudes were believed to lead to a driven work ethic. The American geographer Ellsworth Huntington infamously argued in the early twentieth century that human civilisation attains it highest expression only in regions that are climatically conducive to mental vigour and physical work (Figure 1.2).

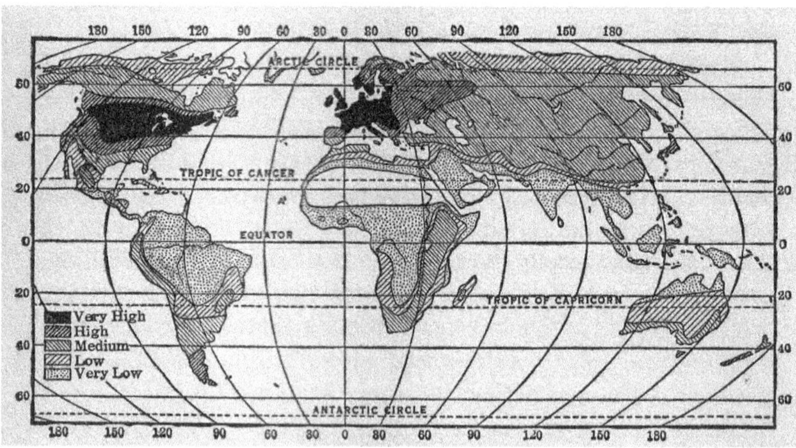

Figure 1.2 Ellsworth Huntington's maps of the distribution of climatic energy (top) and of "civilisation" (bottom).
(*Source*: Huntington, 1915).

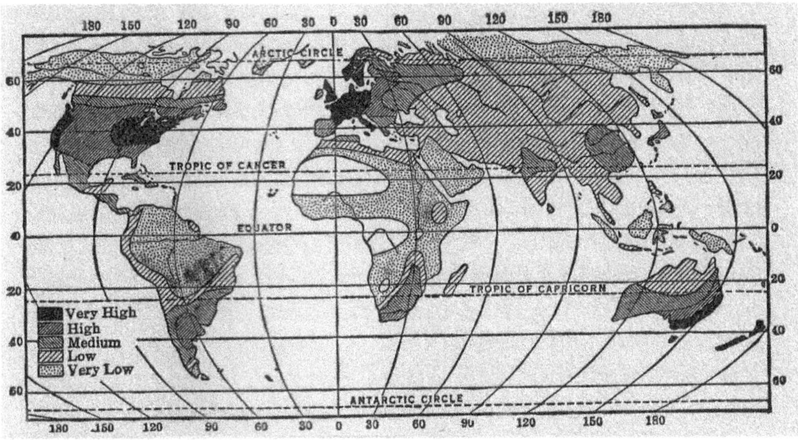

Figure 1.2 Continued

BOX 1.1 ELLEN SEMPLE (1863–1932) – CLIMATIC DETERMINISM

In 1911, the 48-year-old American geographer Ellen Churchill Semple (Figure 1.3) published *Influences of Geographic Environment*, a treatise on what would later be called climatic or, more broadly, environmental, determinism. Semple's book appeared at a time when geography was first emerging as an independent academic discipline in the United States and Semple – then teaching at the University of Chicago – was one of America's first female professional geographers. In 1921 she became the first female president of the Association of American Geographers (and it would be another 40 years before the second female president).

The central idea of *Influences* was that the organisation of human societies is substantively shaped by the environments within which those societies develop, not least by their climate. Semple's work was influenced by her extensive world travels, which included visits to Japan, Korea, China, the Philippines, Java, Ceylon (now Sri Lanka), India, Egypt, Palestine, Lebanon, Turkey, and Europe. This notion of climatic determinism was espoused by many of the relatively small community of professional Anglo-American geographers at the time; its influence on the new discipline of geography was significant.

(*Continued*)

Figure 1.3 Ellen Churchill Semple, c.1914.
(*Source*: University of Kentucky Archives).

Writing a hundred years later, Innes Keighren (2010) explains how Semple's – and other similar ideas – propagated through the early decades of the twentieth century. In the early twentieth century, it was commonplace to talk about "the geographical factor" – including climate – in human affairs. It was part of the attempt to make the discipline relevant to wider society, much as Tim Marshall's popular – if rather superficial – *Prisoners of Geography* (2015) has sought to do in recent times. Keighren also explains why Semple's book was received differently in different places and why it subsequently aroused the passions it did. Even though the form of determinism expressed by Semple and others became discredited later in the twentieth century, not least because of their chauvinist undertones, expressions of climatic determinism continue to be found in recent writings on climate and history, for example Jared Diamond's *Guns, Germs and Steel* (1997) and Geoffrey Parker's *Global Crisis: War, Climate Change and Catastrophe in the Seventeenth Century* (2013).

Along similar deterministic lines, the American public policy professor Solomon Hsiang has claimed in a series of empirical papers written over the past decade that, worldwide, climate is a chief determinant of economic production, human violence, and suicide rates (see Hsiang et al., 2013).

Read:

Keighren, I.M. (2010) *Bringing Geography to Book: Ellen Semple and the Reception of Geographical Knowledge*. London: I B Tauris; Semple, E.C. (1911) *Influences of Geographic Environment on the Basis of Ratzel's System of Anthropo-Geography*. New York: Henry Holt & Co; See also Hulme, M. (2011) Reducing the future to climate: A story of climate determinism and reductionism. *Osiris*. 26(1): 245–266.

In this way, climate has not only been understood as the *maker* of individual and social character but also as the prime *mover* of people. Climate and its variations offer a seemingly powerful and simple explanation for why people move, whether these be Mongol nomads of the sixth century CE driven by climate variations in search of better pastures or – in recent years – Syrian refugees forced to flee a violent civil war "caused" by drought (Selby et al., 2017). Both these dimensions of climate determinism – the making of character and the moving of people – emphasize the power climate is believed to exert over human actions.

One reason climate is so frequently used this way is that it offers an identifiable and apolitical – but often quantifiable – locus of blame. Invoking climate as the "natural" reason for some social ill or economic failure appears to absolve human actors from responsibility. Climate becomes a means of deflecting attention away from political reasons for failure or else of blaming such failure on whatever agents might be held responsible for climate change (see the following section). The relationship between climatic and human agency – the partitioning of responsibility for social and environmental phenomena – is, however, much more nuanced than is usually claimed. For example, the difference between saying that the Syrian civil war that erupted in 2011 was "caused", "affected", "influenced", "shaped" or "conditioned" by climate change is subtle and hard to define and defend. These terms have all been used to describe climate's ostensible causal role in the war and the subsequent flow of refugees out of the country. Yet the Syrian civil war was more a result of the economic liberalisation and ethnic chauvinism of Bashar-al Assad's political regime than it was "caused" by a meteorological drought

(Daoudy, 2020). Its origins should be searched for in antecedent political decisions rather than attributed to climate change.

When climates change

It is not just narratives of climatic blame for social ills or political misjudgements and misdemeanours that are long-standing. So too are human anxieties about a disorderly climate. These anxieties are manifest today in popular descriptions of climate change such as "weather weirding" or climate "breakdown", "chaos" or "crisis". If the idea of climate serves to stop the human world falling apart – as I have argued previously – then it begins to explain why the perception of climates *changing* is so unsettling. It undermines the "trust" people have placed in a stable climate as a reassuring symbol of cosmic, political, social, or ecological order.

The irony is conspicuous. The very idea that cultures have developed to bring order *out* of chaos – the idea of "a climate" – itself now appears to be chaotic. This erosion of the imaginative trust placed in the idea of climate has real material, political, and psychological consequences. Disturbances to the atmosphere, preternatural weather extremes, the apparent disordering of the seasons, the sense of a general "breakdown" in climate . . . these all become portentous events (Daston, 1991). They are not merely manifestations of extreme meteorological phenomena. They are signifiers of a world in crisis. As with early modern Western societies – and in many cultures still today – such events are believed to take on momentous significance beyond themselves.

When climate is believed to be imminently changing, or when the prospect of future change looms large, climate becomes a cultural problem as much as it presents as a physical challenge. When physical climates appear to be changing, the search for convincing explanations for such change becomes pressing. This raises difficult questions. How do people perceive that their climate *is* indeed beginning to change? And what happens to cultural frames of reference and psychological ordering when *perception* of change becomes *conviction*? Have the deities been provoked to punish a wayward humanity? Is an offended God calling for repentance? Has the atmosphere been "prodded and poked" too far through excessive human behaviour for it any longer to yield the climate people expect or require?

This need to explain climatic disorder prompts various lines of reasoning. Within many cosmologies, disruption to settled or ancestral relationships between humans, non-humans, and the spirits or gods is believed to herald adverse consequences for the behaviour of weather. Such beliefs held sway, for example, over many Chinese dynasties. Thus, during the Qing Dynasty, the Shunzhi Emperor in the 1650s – faced with the devastating effects of floods in the capital – could declare that Heaven had brought such disaster because of "Our lack of virtue". He promised to reform himself (Elvin, 1998: 218). Public faith in a reliable climate was undermined by disturbances to the moral-material order. For example, the Abrahamic religions – Judaism, Christianity, and Islam – understand a transcendent and omnipotent God to be the provider of all good things, including the provision of orderly and faithful weather. If climate no longer yields these benefits then this becomes prima facie evidence of the need for repentance and/or supplication. Occasions of collective penitence and prayer – such as national "days of prayer" or **rogation days** – are rooted in the belief that religious rituals have material effects on the weather (Hardwick & Stephens, 2020). For many in the world today, this remains an important belief.

Other supernatural explanations of climatic misadventures may be more sinister, drawing attention to a Manichean struggle between the powers of good and evil. For example, in some European cultures in the early modern period, the devil's work in disrupting the orderly state of climate was believed to have been accomplished through the intermediary of the weather-making witch. And some early colonial writers from the seventeenth century were convinced that the harsh climate they encountered in America was somehow linked to the paganism or witchcraft of the continent's native inhabitants. Their belief was that the coming of Christianity would make the climate more temperate as paganism was eradicated (White, 2015).

There is a more general implication for religious communities of this search for explanation. If deities or spirits are deemed powerful and awesome, then a prerequisite for retaining a beneficent climate is for humans to maintain good and appropriate relations with these supernatural beings. When rituals are neglected or perverted, climates will change. For example, Jerry Jacka observes for the Porgeran people in the mountain valleys of Papua New Guinea how climate change is attributed "to societal breakdown between [the Porgerans] and the rituals oriented toward powerful spirits that control the cosmos" (Jacka, 2009: 206). Similarly, in the highlands of Peru, Catholic farmers attribute climate-related problems such as poor harvests, hailstorms, and sick cattle to the abandonment of *pago a la tierra* – a ritualistic practice that dignifies the Earth as a sacred/social "person" special to God. A recurrent belief across many human cultures is that climate change is caused by a world unravelling spiritually, socially, and ecologically. In her book *Caring for Glaciers Land, Animals, and Humanity in the Himalayas* (2019), Karine Gagne explores this same phenomenon in the Himalayan region of Ladakh, in northwest India.

But just as physical climates change over time, so do theories of climatic causation. Explanatory accounts of *why* climates change do not remain static. As cultures evolve – often in response to experiences of environmental change, cross-cultural encounters, new scientific knowledge, or technological innovation – so too do explanations of climatic change and variability. And there are moments in human history when ideas of climatic causation change particularly rapidly and acquire new cultural authority. For example, the novel idea that climates could change over vast epochs *without* reference to God's immediate agency became widely accepted across the Western world in the middle decades of the nineteenth century. The epistemic conditions for such intellectual novelty were lain down in earlier centuries of European inquiry through the rise of empiricism, scientific instrumentation, and global exploration. Also necessary for the emergence of such new theories of climatic change was a re-imagination of the European idea of time, aided by the work of eighteenth-century geologists such as James Hutton and Georges Cuvier. Rather than 6,000 years of a divinely maintained climate, it became possible to imagine that the Earth may have manifest many large fluctuations in climate during a "deep past" consisting of millions of years. With the realisation of the vast natural forces required for global climate to vacillate through repeated glacial cycles, the search was on to identify the natural causes – beyond divine fiat – of such disruptions.

From this very brief survey of changing cultural explanations for climatic change, the following can be concluded. On the one hand, climates are believed to change due to supernatural entities. These entities enact their agency over climate either directly or else indirectly through natural means. In other words, accounts of supernatural climatic causation may or may not be accompanied by naturalistic explanations. On the other hand, climates are believed to change for *entirely* natural reasons, as in the scientific understanding of glacial cycles. Yet this neat distinction between two seemingly mutually exclusive explanations of climate change – the supernatural and the natural – does not fully reflect how many people commonly think about causation. Human agency is implicated in most *supernatural* accounts of climatic causation in diverse and complex ways. And similar complexities afflict thinking about human agency and *natural* causation. Most cultures have therefore accommodated human agency in *both* supernatural and naturalistic modes of explanation. God does not act independently of human behaviour. Neither is nature unresponsive to human actions.

The question then becomes, "How do different cultures – and different people within particular cultures – apportion responsibility for climatic change between their gods, nature, and themselves?" The boundaries among these different modes of explanation are far from clear, are never static, and are frequently contested. Monotheists believed that it was human wickedness that provoked God to intervene to cause Noah's Flood (see Box 8.1). Aristotle and his disciples believed that human-cleared forests caused the climate of Philippi in ancient Greece to warm. Early European settlers sought to understand the unanticipated harsh and volatile climate of eastern America by reasoning that European climates at similar latitude must have been tempered by millennia of human deforestation, land improvement, and cultivation (Kupperman, 1982). And in post-revolutionary France in the early nineteenth century, the socialist Charles Fourier was convinced there was a decline in the health of planetary climate caused by human greed (Locher & Fressoz, 2012). It is a dangerous form of presentism that thinks that it is only late-modern Westernised cultures that have identified a role for human agency in causing climates to change (Miglietti, 2022).

Supernatural, natural, and human explanations for climate change therefore coexist within and across different cultures. There is an ebb and flow to their respective cultural authority. Nevertheless, the historical evidence would suggest that it is exceptional for humans to think that climates change for either natural or supernatural reasons *alone*. Far more common – and indeed perhaps more necessary – is for people to believe that the performance of climate is tied to the behaviours of morally accountable human actors. For much of the past – and still commonly today – climate and humans have been understood to move together, their agency and fate conjoined through the mediating roles of natural processes and supernatural beings. Thus, English floods and American hurricanes are blamed by early twenty-first-century evangelical pastors on the permissiveness and sinfulness of Western society. Global warming is blamed by eco-socialists on the rampant greed and conspicuous consumption of globalised capitalism. Andean farmers blame capricious weather on their own failure to pacify the Earth spirit. These cases and many more demonstrate the extent to which people – whether moderns or non-moderns, religious and not – regard themselves or others as morally culpable for the disturbed state of their climate.

Placing climate change in history

We are now in a position to see why any exploration of the idea of climate change should start with a look into the past. The changing climate has frequently been woven into historical accounts of the fate of people, societies, and nations. Similarly, in the contemporary imagination, climate change is deeply implicated in the many stories told about how the human future will unfold. (We will see some examples of this later, especially in Chapters 7 and 10.) The role of the climate historian as storyteller, sage, and prophet therefore becomes important. As Martin Mahony and Sam Randalls explain in the introduction to their recent work, *Weather, Climate and the Geographical Imagination*, climate

> plays a powerful, sovereign role in its imagined capacity to fundamentally shape human geographies of violence, economic prosperity, and environmental vulnerability. Understanding this lineage not only

> of climatic determinism but of climatic expectation more broadly is a critical historical task with urgent contemporary resonances.
> (Mahony & Randalls, 2020: 4)

Lessons can be learned from studying climate's *historiography*, the systematic and critical analysis of texts that have been written about the role of climate in history. Cultural geographers and environmental historians, literary critics, and climate historiographers can reveal how climate has been written into – or out of – human history. Mark Carey's review of climate's historiography shows how scholarship on the cultural histories of climate and its changes may be crucial for preventing understandings and fears of future climate change from becoming too presentist (Carey, 2012). Carey shows that, perhaps surprisingly, there is much more climate historical scholarship dealing with the period before 1900 than for the twentieth century. Climate historians such as Jean Grove (the little ice age in Europe), Richard Grove (the colonised tropics), William Meyer (nineteenth-century America), David Zhang (Ming and Qing dynasties in China), and Sam White (the Ottoman Empire) have led the way in this climate story-telling.

The first international conference on climate and history was held not much more than four decades ago, at the University of East Anglia in the UK in July 1979. It was convened by one of the academic founders of the sub-discipline of historical climatology, Professor Hubert Lamb (1913–1997). Although fascinated by history, Lamb graduated from the University of Cambridge with a degree in geography before entering the British meteorological service. Along with Emmanuel Roy Ladurie, Christian Pfister, and Reid Bryson, Lamb was one of the first prolific scholars of history to draw attention to the importance for human history of climate variations on the time-scales of decades, centuries, and millennia. Lamb was an early pioneer of extracting climate *from* history; he used historical sources to reconstruct climates of the past millennium. But he also wrote climate *into* history; he offered accounts of how climate can be an agent of historical change (Lamb, 1982).

Lamb's work demonstrated how climate historians need to be faithful to whatever historical traces and scientific and cultural understandings of past climate-society interactions can be retrieved. Physical geographers and paleoclimatologists can excavate and interpret material traces of past climates, whether from tree or coral rings, pollen and algae in lake sediments, air bubbles in ice cores, volcanic tephra (rock fragments), or speleothems (cave deposits). From these evidences, variations in the physical behaviour of the atmosphere can be reconstructed. Social and historical geographers and environmental historians – relying upon material human artefacts and documentary sources – have a different task. They seek to develop accounts of how such past fluctuations and changes in climates may have interacted with the development of human societies.

For example, the historian Dagmar Degroot explains why the resilience of societies to climate fluctuations and change is a dynamic and highly contingent process (also see Box 1.2 and the work of Georgina Endfield). In his history of climate change and the Dutch Republic in the seventeenth century – *The Frigid Golden Age* (2018) – Degroot shows how the climatic fluctuations of these middle decades of the "little ice age" had very different repercussions for societies. He argues that Dutch and Japanese societies prospered in these years, even while other societies faced severe economic and social challenges. But the reasons for their respective prospering were deeply contingent upon the antecedent political and cultural conditions of these two nations. The impacts of climate change in the seventeenth century on human societies could not simply be determined through correlations with a quantitative climatic index.

BOX 1.2 GEORGINA ENDFIELD (B.1969) – CLIMATE, HISTORY, AND SOCIETY

Georgina Endfield, Professor of Environmental History at the University of Liverpool, is a historical geographer who has made significant contributions to understanding the complex relationships between climate, weather, and societies in the past. She gained both her BSc (geography) and MSc (archaeology) from the University of Liverpool, while her PhD was awarded by the University of Sheffield for a thesis titled "Social and Environmental Change in Colonial Michoacan, West Central Mexico". Her PhD thesis was a study of climate variability and societal resilience in colonial Mexico in the eighteenth and nineteenth centuries and led to her later monograph *Climate and Society in Colonial Mexico: A Study in Vulnerability* (2008).

Endfield's work embraces climatic history and historical climatology, human responses to unusual or extreme weather events in history, and the links between climate and the healthiness of places. She argues for a deeply contingent relationship between climate and society. Through her detailed, archive-based studies of the many ways in which individuals, communities, and societies remain resilient in the face of climate change, Endfield challenges simplistic climate-driven "civilization collapse" narratives that have become popular in recent years. Her work shows that the experience of climate variability, extreme weather, and

(Continued)

weather-related events and crises can indeed challenge societal resilience. But it can also increase opportunities for learning and innovation, extending the repertoire of collective adaptive responses. Through historical examples extending from colonial Mexico, to southern Africa in the nineteenth century and the United Kingdom over the past 500 years, Endfield elucidates the ways in which societies develop strategies to deal with climatic perturbations at different scales. Social breakdown and collapse are not an inevitable result of climatic change or extreme variability. Her more recent work on the history of English climate over the past half-millennium has also drawn attention the important role played by personal and cultural memory in the social mediation of climatic extremes.

Read:

Endfield, G.H. (2008) *Climate and Society in Colonial Mexico: A Study in Vulnerability*. Oxford: Blackwell; Endfield, G.H. and Veale, L. (eds.) (2017) *Cultural Histories, Memories and Extreme Weather: A Historical Geography Perspective*. Abingdon: Routledge.

Which hints at a further, more challenging task for the climate historian. This is to recognise the power of the human imagination in conditioning how the material interactions between climates and societies are experienced. What different conceptions of climate have prevailed in past historical cultures? What notions of time and change shaped earlier cultural interpretations of extremes of weather? These cultural heuristics and imaginative devices leave few or no material traces for the scientist or archaeologist to unearth. One is left with using literary or other more speculative cultural analyses to enter into the mind of early and pre-moderns (Barnett, 2019). As we have seen previously, past societies attributed different forms of agency to climate and understood the signifiers and causes of climate change in very different ways. Recent work in **material eco-criticism** emphasises the co-constitution of climate and culture, as for example exemplified in the work of literary critics Adeline Johns-Putra and Sophie Chiari (see Chapter 7).

Chiari, for example, has interrogated at length Shakespeare's climatic imagination and how he approached the sky as a theatrical element of his plays. To Shakespeare and his contemporaries in late sixteenth-century

England, climate was not something that could be reduced to numbers. The vicissitudes of climate were the vehicle for revealing fate and for conveying judgement to individuals and societies alike. Chiari argues that climate was a framing device for Shakespeare, which offered a coherence to his playtexts and which provided his audiences with natural, elemental, and cosmic backgrounds. "Influenced by Greek medical and meteorological thinking", Chiari argues, "[Shakespeare's] plays allow men and women to undergo a sensory and sensitive experience of the weather and of sky-related phenomena, one that could free them from dread and which could make them dream of a 'brave new world' " (2019: 7).

Past debates about climate and its role in human life have always been about more than merely the physical weather of places. They are always wrapped up with arguments about power, religion, knowledge, economics, values, and identities. (We will see this applies equally to future climate change, Chapter 10.) Constructing accounts of earlier understandings of how climates and cultures interact in the human imagination – whether for the individual settler, the aboriginal community, the economic migrant or the political or religious elite – is therefore a difficult task for the historian. Nevertheless, it is an important one. The environmental historian Lawrence Culver advocates thus (2014: 317): "In seeing climate through culture, [the historian] can enrich environmental history, enlarge national and transnational histories, and perhaps even inform our own future as well". What is learnt from environmental and cultural historians and historical geographers is that people in the past reacted to climatic extremes and perturbations in a multitude of ways. Formal codified knowledge – what today is defined as scientific knowledge of climate – was lacking. But different cultures developed their own heuristics for reading the skies and for deciding how to act in the face of climatic uncertainty (Jankovic, 2000). Different conceptions of climate in the past functioned as valuable cultural resources, as much as they described a fixed material reality.

Chapter summary

This chapter advances the argument that to understand the idea of climate change and its power in today's world, it is necessary to start with geography and history rather than with science and technology. The idea of climate emerges variously in historical cultures as a means to order and regulate the human experience of turbulent and unstable weather. A historical-geographical approach to the study of climate change, as exemplified in this chapter, therefore points to the importance of the cultural imagination for fully understanding climate and its changes, causes, and consequences.

These cultural readings of climate and its changes pre-date, by many centuries, the scientific accounts of the atmosphere's changing behaviour that will be the subject of the next chapter. Climate change is not – and never has been – only a physical phenomenon. Rather than signifying a universal physical reality, it is more fruitful to understand climate change as an idea that accommodates multiple meanings. People's interpretation of climate change is deeply shaped by their histories, cultures, materialities, relationships, and imaginations – how and where they live, think, and act. This leads inevitably to a political struggle for meaning.

Opening up different histories and geographies of climate is necessarily subversive – in the same sense as were French social theorist Michel Foucault's archaeologies and genealogies of madness, punishment, and sexuality when written in the 1960s and 1970s (see Rabinow, 1991). Or perhaps I can put it this way. Histories and geographies of climate are subversive of claims about the *matter-of-factness* of what is today known scientifically about climate change. One cannot extract the meaning of climate change from a scientific text. Neither can one discern it simply through the experience of weather, however extreme or unusual. And as the six chapters in Section 2 will later show, different cultural beliefs, human values, political commitments, and collective visions of the future *continue* to guide interpretations of new scientific evidence of a changing climate. These meanings are dictated neither by science nor by scientists.

Note

1 Interestingly, this was also the device used in the late eighteenth century by Sir Joseph Banks when communicating to his government in London what the unfamiliar climate of Botany Bay in eastern Australia was like. Banks compared the climate of what today we call "Sydney" to the climate of Toulouse in the south of France, in the process transposing by 10° of latitude these two hemispheric mirror-imaged climates (Mercer, 2020).

Further Reading

Two good places to start reading about climate and culture are **Sarah Strauss** and **Ben Orlove's** edited book *Weather, Climate and Culture* (Berg Press, 2003) and

Wolfgang Behringer's *A Cultural History of Climate* (Polity, 2010). Both books explore societies' relationships with weather and climate and reveal the complex influences on how people interpret meteorological phenomena. One of **Mike Hulme's** previous books – *Weathered: A Cultural Geography of Climate* (SAGE, 2016) – also explores how different cultures make sense of their weather and the different cultural functions served by the idea of climate. Also on offer is **Giuseppe Feola** and colleagues' edited volume *Climate and Culture: Multidisciplinary Perspectives on a Warming World* (Cambridge University Press, 2019), which contains a series of geographical case studies of how climate change can be understood through cultural investigation. There are now quite a few substantial books by historians examining interactions between climates and societies over the past millennium. Two of the best are **Dagmar Degroot's** *The Frigid Golden Age: Climate Change, the Little Ice Age, and the Dutch Republic, 1560–1720* (Cambridge University Press, 2018) and **William Meyer's** *Americans and Their Weather: A History* (Oxford University Press, 2nd edn. 2014). **Sophie Chiari's** *Shakespeare's Representation of Weather, Climate and Environment: The Early Modern 'Fated Sky'* (University of Edinburgh Press, 2018) is a unique exploration of how the idea of climate change was imagined by Shakespeare in the culture of early modern England. **Sara Miglietti** does something similar in *The Empire of Climate: Early Modern Climate Theories and the Problem of Human Agency* (Cambridge University Press, 2022) but with respect to the political nature of climate change in early modern Europe. Finally, **Sam White, Christian Pfisterbv**, and **Franz Mauelshagen** have edited a valuable collection of essays – *The Palgrave Handbook of Climate History* (Palgrave Macmillan, 2018) – about the reconstruction of climatic history and about the historical impacts of climate change on societies around the world.

QUESTIONS FOR CLASS DISCUSSION OR ASSESSMENT

Q1.1: In what ways have ideas of stability and change been central to historical understandings of climate?

Q1.2: In what sense – and under what conditions – might it be legitimate to describe climate as "a cause" of social phenomena?

Q1.3: Does the idea of climate change only make sense to people through their lived experience of place? How might the advent of the internet and social media have "*dis*-placed" people's imaginative engagement with the idea of climate change?

Q1.4: Why does a change in the climate – whether a real or perceived change – so frequently raise questions about the moral failures and failings of human societies and individuals?

Q1.5: Is Dipesh Chakrabarty (2009: 201) correct to claim that "anthropogenic explanations of climate change spell the collapse of the age-old humanist distinction between natural history and human history"?

2

CLIMATE CHANGE AND SCIENCE

. . . Climate change quantified . . .

Introduction

Over the past two centuries the ascendency of science for providing the basis of much human knowledge – a key feature of the narrative of modernity – has changed the way in which climate is understood across many cultures. For greater numbers of people than previously, climate has come to be understood in strictly physical and explicitly quantitative terms. This scientisation of climate has accelerated over the past 70 years and carries with it the promise of quantitative climate prediction. Two factors were especially important in enabling this change in thinking. First were the transnational scientific networks established in the nineteenth century by European imperial powers. And second – and more decisively – was the influence of the Cold War on the emergence of a new "**infrastructural globalism**" (Edwards, 2006). By the end of the twentieth century, Western science had reconceived the idea of climate as a manifestation of a deeply interconnected physical Earth System. Climates were no longer local, multiple, and cultural; climate had been *made* global, singular and scientific.

This re-conception of climate as something accessible to science led to a further change. There was a growing conviction during the latter decades of the twentieth century – first among scientists, then among publics – that aggregated human activities could substantially change the dynamics of this global climate and indeed were manifestly doing so. Climate was not just global. It was now also human-shaped. These twin developments were not coincidental. The political powers, technologies, and networks that were fuelling the human activities that were transforming the world's climate were the very same ones that were enabling scientific inquiry to reveal the impacts of these activities on the climate. The fossil-fuelled economies of the nineteenth century set in motion the anthropogenic modification of the world's climate. Yet these were the very same political economies that first inspired – and then equipped and enabled – new scientific infrastructures and institutions that were eventually to detect the effects of this modification. This chapter explores how these new conceptions of climate and climate *change* emerged from developments in scientific practice.

Scientific knowledge is inherently situated, a claim that is as true for the new sciences of climate emerging during the nineteenth and twentieth centuries as for any other science. Furthermore, the science of climate developed within particular political cultures and received the benefits of the patronage of these cultures. This chapter is structured around these two arguments. The science of climate change – whether manifest in reconstructions of global temperature, the development of climate models, or the knowledge claims of the Intergovernmental Panel on Climate Change (IPCC) – always represents a view from somewhere. Place matters. And, second, the science of climate – or what has come to be known as Earth System science and in which today's epistemic claims about future climate change are rooted – has its own history. It is one closely bound up with the history of geopolitical power. Politics also matters. To develop these two arguments, I draw upon the work of *historians of science, science and technology studies scholars, critical physical geographers,* and *political geographers.* Together, these various strands of thought and analysis comprise the sub-discipline of geographies of science (Mahony & Hulme, 2018).

***

The chapter begins by showing how the supplanting by science of older culturally conditioned ideas of climate – those introduced in Chapter 1 – can only be understood through careful attention to the spatial, social, and political relations of scientific practice. To illustrate this point further, I then consider the importance of political patronage for climate science across three different eras: The European imperialism of the nineteenth century; the militarism of the Cold War; and the internationalism of the United Nations. The scientific claims about climate change put forward since 1990 by the United Nation's IPCC have proved epistemically and politically powerful. But the role of the nation state as the incubator or interpreter of certain aspects of climate science has never been fully replaced by the United Nations. I end the chapter on this point. Even in an era of extensive international and online scientific cooperation, "place" – as manifest through state politics and cultural norms – retains its influence on the making and transmitting of scientific knowledge about climate change.

Place matters

Science is an inherently geographical method of knowledge-making. That is, scientific activity occurs *in* places. Scientific instrumentation and measurement, the initiation and testing of scientific theories through field observations, laboratory experimentation, or computer modelling, the pursuit of scientific consensus – these are all social practices of science that are bound up with place-based cultural norms and expressions of political power. The conduct of science is shaped by these contextual factors, which vary across time and space (Livingstone, 2003).

We can illustrate this through the example of the Swedish glaciologist Hans Wilhelmsson Ahlmann (1889–1974). The inter-war decades of the 1920s and 1930s witnessed growing awareness among some scientists and many local observers that climates in parts of the world were appearing to warm. Thus, in the early 1930s a professional American meteorologist could note,

> The phase of . . . climate that is attracting attention at the present time is . . . an apparent longer-time change to cool periods that seem to be less frequent and of shorter duration, and warm periods that are more pronounced and persistent.
>
> (Kincer, 1933: 251)

This climatic warming was particularly noteworthy in the Arctic North and a leading proponent of this view was Ahlmann. Having led a series of field expeditions in the 1920s and 1930s to monitor and study glaciers in Greenland and Iceland, Ahlmann subsequently promoted widely his argument about "polar warming" (Sörlin, 2011). In the late 1940s/early 1950s this was well before any wider sense of a global change in climate had taken hold.

Ahlmann's attention to careful local observations of slowly changing physical properties of glaciers is a classic example of a field science. It was systematic observations of a changing environment *in places* – combining the insights gained from scientific instrumentation with local knowledge of retreating glaciers held by settled Indigenous inhabitants – that yielded evidence of a changing climate. Ahlmann was committed to a distinctive form of field science that was obviously located among these Arctic glaciers. For him, the glacier became what sociologist of science Thomas Gieryn (2006) has called a "truth-spot", a specific place that lends credibility to beliefs and claims about the natural and social world. Ahlmann practiced a distinctly "Nordicist" culture of field glaciology (Sörlin, 2011) that allowed these Greenlandic and Icelandic glaciers to yield authoritative empirical knowledge of changes in climate, whether these changes were of natural or anthropogenic origin.

And yet Ahlmann's approach to the scientific study of a changing climate was not universally shared in the postwar period. Or at least one might say that there were other *places* in which scientific knowledge about climate change were being made. This knowledge too bore the specific marks of other "truth spots". These included, for example, the Mauna Loa Observatory in Hawaii established by Charles Keeling in 1957 to measure atmospheric concentrations of carbon dioxide. Or the Geophysical Fluid Dynamics Laboratory at Princeton University where Syukuro Manabe pioneered in the

1970s the three-dimensional simulation of global climate using computer models. Or the Climatic Research Unit in the School of Environmental Sciences at the University of East Anglia in Norwich, which produced in the 1980s the first comprehensive and systematic compilation of global surface air temperature variations (Figure 2.1). Each of these places mattered

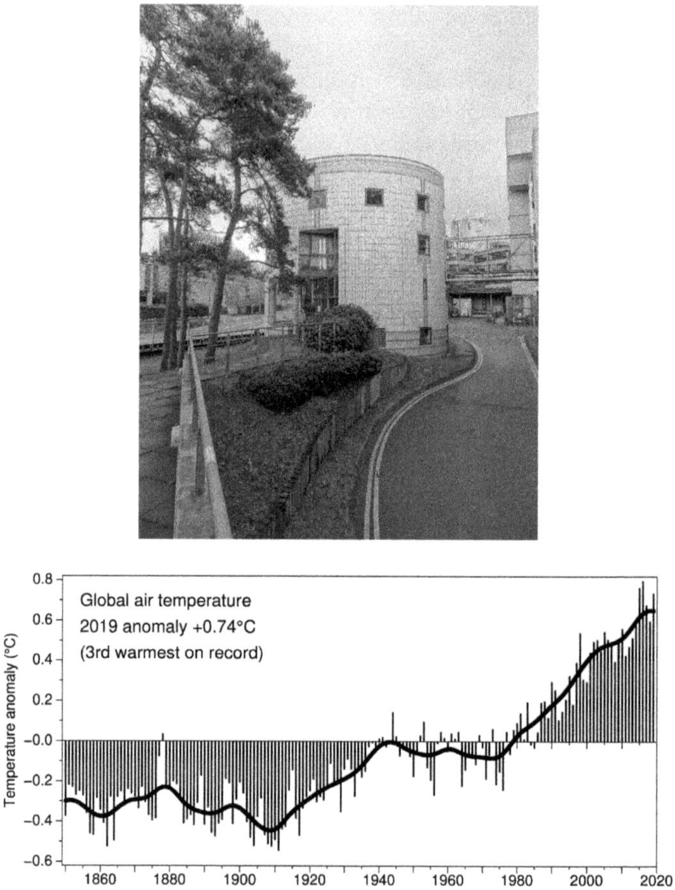

Figure 2.1 (Top) The Climatic Research Unit at the University of East Anglia, UK, where in the early 1980s the first comprehensive and systematic empirical record of global surface air temperature variation was compiled. (Bottom) The HadCRUT global temperature curve, 1850–2019; anomalies are with respect to 1961–1990 baseline.

(*Sources*: Author, top and Climatic Research Unit, bottom).

in material ways for the form and credibility of scientific knowledge that was put into wider circulation during these decades. This credibility was achieved through, for example, establishing evidential standards, forging new ways of relating theory to data, or quality-controlling empirical meteorological data.

Today's universal claims made by climate scientists about the current and future condition of the world's climate are very familiar to us. So, it is easy to miss this point: *All* scientific knowledge about climate change has its geography. It is made in some places and not made in others. For example, the computer models that underlie many scientific claims about climate change are funded, designed, and operated in specific places with their own political and cultural histories (see Box 2.2 later in the chapter for an example of this). But so too are these models utterly dependent for their design, calibration, and testing upon theories and empirical data that first emerge in specific places. Similarly, climatic concepts or metaphors – which may seem timeless and placeless, essential rather than socially constructed – have their own geographical origins. They are created by specific people in certain places at particular moments. Thus, the christening of the episodic warm eastern Pacific Ocean current El Niño (Grove & Adamson, 2018), the naming of the Leeuwin Current off western Australia (Morgan, 2020), or the metaphors of Gaia (Lovelock, 1979) and **tipping points** (Russill, 2015) . . . these are all examples of climatic "discovery" and ordering that can be traced to a particular place, period, or person. The spatial contingency of scientific knowledge is the central insight offered by historians, anthropologists, and geographers of science such as Steven Shapin, Bruno Latour, and David Livingstone.

But this observation presents us with a paradox. The practices of climate science are intensely local – observing glacier movements on mountains, measuring the atmosphere at meteorological observatories, operating climate models through networked computers, coring trees in boreal forests. Place-based science relies on the local confluence of personalities, politics, and serendipities, as elegantly outlined in Sarah Dry's book *Waters of the World* (see Box 2.1). And yet the scientific claims about a changing climate that result from such practices appear to achieve global reach and universal validity. To resolve this paradox requires an appreciation of the international networks of practice and circulation enabled by powerful political patronage. Only thus can such local knowledge be transformed into "global kinds of knowledge" (Hulme, 2010) and subsequently secure the epistemic and political influence that comes from this transformation.

BOX 2.1 THE MAKING OF CLIMATE SCIENCE – *WATERS OF THE WORLD* BY SARAH DRY

Science is an immensely powerful cultural pursuit, a process of inquiry into the functioning of the material worlds around us, within us, and beyond us. It is a process that is both individual and collective, both personal and social. The particular disciplinary formations and traditions within science are also inventions of human societies. They are contingent upon people, places, patronage, and politics. This is very much the case for what today is called the discipline of climate science, or Earth System science. Science historian Sarah Dry offers an account of the contingencies involved in the making of climate science in her book *Waters of the World* (2019). Dry shows how people, places, and politics interacted in complex and sometime unforeseen ways to create a new understanding of climate as something that emerged from globally connected physical processes and that was intelligible to science. Her book is subtitled "The story of scientists who unravelled the mysteries of our seas, glaciers, and atmosphere – and made the planet whole". The contingencies that Dry explicates illustrate why science – and climate science in particular – has its particular geographies.

Dry's account of the science of climate change is important for two reasons. First, it challenges what she describes as "a larger attempt by climate scientists to tell a singular history of [what is] a heterogeneous science", an attempt that risks obscuring the multiple and contingent ways in which climate and its changes made are known. For Dry, it is crucial to show how climate science has no single or inevitable lineage. Human *ways* of knowing the climate are subject to evolutionary processes of change (Renn, 2020) – as too are physical climates themselves. Second, *Waters of the World* tells its story through the lives, successes, and failures of six individual scientists over the past 150 years – John Tyndall, Charles Piazzi-Smyth, Gilbert Walker, Joanne Simpson, Henry Strommel, and Willi Dansgaard. Dry is therefore able to offer historically and geographically rich insights into the sometimes febrile and chaotic processes by which scientific knowledge of climate gets made: The personal disappointments and dead ends; the contingencies of patronage and funding, education, and training; family relationships and illnesses; and the sometime fortuitous access to instruments, institutions, scholarly networks, and other people's labour. As Dry concludes, "Global visions are necessarily made up of unglobal things – individuals, places, moments in time". Today's scientific knowledge of a changing global climate has emerged, remarkably, from such chaotic and unplanned activities.

Read:

Dry, S. (2019) *Waters of the World: The Story of the Scientists who Unravelled the Mysteries of our Seas, Glaciers and Atmosphere – and Made the Planet Whole*. Chicago: University of Chicago Press; See also: Hulme, M. (2010) Problems with making and governing global kinds of knowledge. *Global Environmental Change*. 20(4): 558–564.

There is another sense in which scientifically fashioned climatic knowledge manages to escape its place-bound origins. It is not just that scientific knowledge about climate change seemingly travels freely across borders – beyond its places of making – thereby becoming globally salient and authoritative. It is also that climate scientists are able to reconceive the idea of climate as being multi-scalar. Climate change becomes an idea that is meaningful across different scales of time and space. Through scientific knowledge it is valid to describe decadal fluctuations in climate of a specific locality – such as the Swiss Alps or the island of Barbados – but also to speak of planetary oscillations in climate over millennia. Scientific knowledge is made in places. But by making this knowledge commensurable – by describing climatic phenomena using the language of mathematics – science also makes things scalable.

Take the example of climatic information extracted from trees, the science of dendroclimatology. Past climatic fluctuations of a specific location can be extracted from a single tree's growth rings (Figure 2.2), perhaps extending back over hundreds of years. This transcends the limitations of time. And by aggregating such individual sequences acquired from a judicious selection of geographically dispersed trees, an estimate of hemispheric or global climate can be derived. This transcends the limitations of space. Or for another example of how the scientific conception of climate escapes the limitations of place or fixed scale consider the language of "downscaling". Here, panoptic, global kinds of climatic knowledge – for example knowledge derived from model simulations of future of climate – are re-expressed at local or human scales through practices of downscaling. For Schneider and Walsh (2019) this practiced is better described by the metaphor of "zoom": Climatic descriptions

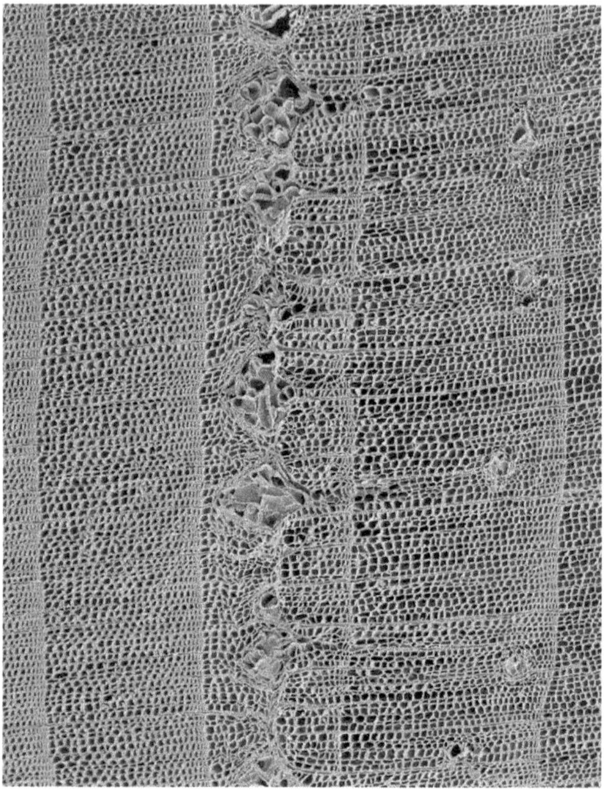

Figure 2.2 Thin section of a tree-ring sample shows a frost ring in 536 CE (Büntgen, 2019; courtesy of P.J. Krusic). This anatomical disturbance was likely caused by a severe cold spell in the early summer of this year, following a large volcanic eruption which marked the onset of the Late Antiquity Little Ice Age (Büntgen et al., 2016). The image contains a sequence of four complete annual tree rings (from left to right): Years 535 CE (wide), 536 (narrow: cell collapse), 537 (narrow: cell recovery), and 538 (wide).

appear effortlessly to transcend the limitations of scale by zooming in and out, as with a photographic lens.

In Chapter 6 we will see some of the barriers to these deceptively easy movements of the scientific notion of climate. These are frictions to movement and assimilation that emerge from different cultures with their own distinct ways of knowing the world. But for now, in the three subsections that follow, I trace the emergence and central importance of *climate infrastructures* and *practices* that are enabled through political patronage for achieving such transcendent knowledge. These include, for example, global instrumental networks (both surface and satellite), global climate models, and international knowledge assessments. Using the work of Deborah Cohn, Martin Mahony, and Paul Edwards, among others, I show how new infrastructures and their associated epistemic practices reinvented climate as something discoverable – and hence predictable – by science.

Imperial and colonial science

The European empires of the nineteenth century offered powerful patronage to the new sciences of meteorological observation, systematisation, interpretation, and communication that began to forge new scientific understandings of climate. These sciences were, on the one hand, enabled by the new imperial infrastructures that were being laid down around the world – physical observatories, accredited observers, telegraphic networks, professional bureaucracies. On the other, they were called upon to serve the new commercial and political interests of these imperial powers. Today's scientific conception of climate as global can be traced back to these new scientific impulses set in motion through imperial patronage. One should not see this development as inevitable. A different unfolding of scientific engagement with the idea of climate was possible. Nevertheless, climate became a globally interconnected idea *conceptually* at the same time as empires and colonies were themselves becoming connected *materially* through the infrastructures of transglobal trade, mobility, and communication. As explained by Mahony and Endfield (2018: 11), the new scientific practices of climatic knowledge-making were "developed in the cultural and epistemological contact zones of the European empires".

The nineteenth century saw new social technologies emerge in Western cultures, which not only enlarged the scale of the human imagination. They also offered new ways of connecting things across scales. Thus, the daily or weekly newspaper became important for creating new national modes of thought; North American railroads provided a new way of thinking about distance and space; and the literary novel offered new expressions of the human imagination that escaped the scalar limitations of the everyday. In similar fashion, the new sites, scales, and connections that European empires offered to science in the nineteenth century were formative in the new science of climatology. These empirical projects of meteorological data gathering and sharing were essential in the iterative process whereby the scientific idea of climate expanded in scale and complexity. Scientists began to argue that atmospheric and oceanic phenomena were connected on ever larger scales and used this assertion to access imperial resources to gather data on unprecedentedly vast scales. The outcomes of this new work of data harvesting and statistical correlation could then be used to justify the initial assertion.

For example, some of the earliest data-gathering activities took place through specialized military weather offices in European colonies and through the trading posts of private commercial initiatives such as the Dutch and British East India Companies. The environmental historian Richard Grove claimed that the year 1791 represents

> the first occasion on which weather and agrarian observations made by scientific observers and others in the tropics were sufficiently elaborate and sufficiently coordinated . . . for some of the first speculations to be firmly made *about global rather than regional climatic events.*
>
> (Grove, 1998: 302; emphasis added)

It was certainly the case that by the early nineteenth century knowledge was circulating through colonial networks about simultaneous droughts afflicting India, the Caribbean, and the white settler colonies in Australia.

The expansion of Europe's continental and maritime empires during the latter decades of the nineteenth century saw the establishment of ever larger networks of formalised weather observations. Deborah Coen shows how in the mid-nineteenth century "the emperor's ministers in Vienna" – the seat of the Habsburg Empire – "saw good reasons, both practical and ideological, to support the study of climate down to the details of the smallest scales" (Coen, 2018: 15). Because the Habsburg Empire extended across large territories in central and eastern Europe, it allowed scientists to develop new investigative tools for studying different atmospheric regimes. New cartographical devices were pioneered for mapping the climates thus conjured into existence. The patronage of the Empire – its need for new narratives of coherence and connectivity across widely diverse landscapes and ethnicities – brought a new science of climate into being. It was a science born of political necessity, argues Coen, "a continental-imperial science of 'regionalisation' with a global vision" (2011: 45).

Similarly, the environmental conditions of the Habsburg Empire stimulated new ways of thinking about *changes* in climate. These included associations among forest clearance, swamp drainage, and climate change and the mediated evidence of climate change available through the new technologies of photography (Figure 2.3). The German geographer Eduard Brückner (1862–1927) – who ended his career at the University of Vienna as the Habsburg Empire drew to a close in the early twentieth century – drew upon river and lake fluctuations, grape harvest dates, grain yields, and glacier movements across central Europe to offer bold prognostications about a 35-year temporal cyclicity in continental climates.

Figure 2.3 Nineteenth-century photographs such as this one of Carls Eisfeld in the Austrian Alps, taken by the Austrian geographer Friedrich Simony (1813–1896), contributed evidence of recent glacier retreat related to anthropogenic climate change. Simony's images of glaciers were of wide scientific and popular interest in his day.
(*Source*: Universitätsbibliothek Wien, Teilnachlass Friedrich Simony).

Central to this new science of an enlarged – but multi-scalar – idea of climate were the regional and national weather services founded in many countries. For example, one can trace the chronology of the unfolding "scramble for Africa" among the European powers in the latter decades of the nineteenth century simply by tracking the commencement dates of the earliest systematised meteorological readings recorded at locations across the continent. The still expanding maritime British Empire facilitated a further extension of the climatic sciences into new lands, islands, and oceans that became sites of essential data collection. One beneficiary of such data was Sir Gilbert Walker (1868–1958), director-general of observatories in India between 1903 and 1920. Using data collected from across Britain's tropical empire – and telegraphed to his offices in Delhi – Walker was one of the first to establish statistical connections between climatic oscillations on a quasi-global scale.

Standardisation – of instruments, observing practices, measurement units – is a canonical practice in the maturation of many sciences. With the increasing – and increasingly rapid – exchange of such weather data via telegraphy, the imperial powers sought also to standardize meteorological measurements.

The first meeting of the International Meteorological Congress (IMC) – a forerunner of today's UN World Meteorological Organisation (WMO) – was held in Habsburgian Vienna in 1873. The IMC desired to ease the exchange of weather information across national borders. The Congress's first president, the Dutch meteorologist C.H.D Buys Ballot, summarised its thinking thus[1]: "It is elementary to have a worldwide network of meteorological observations, free exchange of observations between nations and international agreement on standardized observation methods and units in order to be able to compare these observations".

Similarly, the International Cloud Atlas, first published in 1896 under the eyes of Hugo Hildebrand Hildebrandsson, Albert Riggenbach, and Léon Teisserenc de Bort from the IMC, "had the crisp mark of imperial power on it, the confidence to mark the skies with order as the railways and telegraph wires had marked the land" (Dry, 2019: 111). The ambition of the Atlas was to cover the world, with a view to promoting "inquiries into the forms and motions of clouds by means of concerted observations at the various institutes and observatories of the globe" (Hildebrandsson & de Bort, 1896; quoted in *loc. cit.*). Here was a specific example of how the new science of climate was enabled by networks of imperial power. In time, it was to offer back to its patrons new knowledge that would reinforce imperial projects of trade, settlement, and tropical agriculture.

Climatic imperialism reached its apogee in the case of the British Empire. Britain's meteorological observatories alone – extending along its maritime trading routes and settler colonies – offered a quasi-global scientific network. Britain therefore cultivated an *imperial* equivalent to the meteorological *internationalism* being developed by the IMC. By embracing tropical and austral climates – as well as temperate ones, thus creating an "empire of all climates" (Mahony, 2016a) – British empire meteorology was able to offer correctives to an overly "northern" scientific perspective on climate. Through a series of three imperial meteorology conferences in the inter-war period – held in 1919, 1929, and 1935 – Britain's imperial scientists created their own earlier version of Edwards' "infrastructural globalism" and scientific internationalism. "The Empire was thus a convenient shortcut to a truly 'global' science, while meteorology itself emerged as a potentially powerful new resource as aviation and agricultural developmentalism took hold" (Mahony, 2016a: 29).

Europe's empires enfolded the world with observatories, scientific bureaucracies, and communication technologies. These contributed significantly to new scientific conceptions of climate during the imperial era. Older ways of knowing climate through human subjectivities and cultures in places began to be replaced in the Western world by these new sciences. The geographical imaginary of climate was enlarged among Western elites in remarkable ways. New, unforeseen connections between distant weather phenomena were drawn. At the turn of the century, Gilbert Walker was so bold as to lay claim to something he called "world weather". The emerging idea of climate as *global* can therefore be understood as an intellectual product of European imperialism. Yet ironically much of this new climatic knowledge was gathered and deployed in the service of distinctly *local* projects of military, commercial, and infrastructural significance. Even if *climates* extended beyond nation and empire – epitomised in A.J. Herbertson's "natural climate regions" (Herbertson, 1905) – climatic *resources* were still the preserve of state or imperial control.

Cold war climates

The First World War washed away the Habsburg and Ottoman Empires – and destroyed, temporarily, German and Russian imperial dreams. And the Second World War accelerated the slow decline of the remaining European imperial powers: Britain, France, The Netherlands, Spain, Belgium, and Portugal. Yet the patronage of Europe's empires enjoyed by the scientists of the nineteenth century studying climate was replaced in the new geopolitical world of the Cold War by the patronage of the American – and to a lesser degree the Soviet – military. In the decades following 1945, scientific investigation into climate trends, processes, and predictions grew in capacity, prestige, and influence under the umbrella of the world's two superpowers: The USA and the Soviet Union. The new discipline of dynamical "climate science" – supplanting that of "statistical climatology" (see Chapter 1) – would take shape in the new politics of the Cold War.

The military demands of the Second World War had greatly expanded the intrusion of the state into scientific research – especially in the USA. It seemed natural therefore that in the new conditions of Cold War rivalry laid down in the early 1950s, the state would continue – indeed, expand – its support of science and technology. There was a new optimism about what science could achieve, what Erickson et al. (2013) have termed "Cold War rationality". Extravagant new state expenditures were authorised by both superpowers and new and expanded roles and expectations for science and technology abounded. Historians of science have shown the powerful influence of the Cold War on disciplines ranging from physics, oceanography, and the environmental sciences to molecular biology and psychology.

The antagonism between the two superpowers and their contrasting ideologies constituted a basic category for the understanding of reality. Military competition on a global scale animated all layers of culture. The emerging threat of thermonuclear war was particularly significant. For political and military leaders to understand the full range of environmental and social effects from an exchange of multiple nuclear warheads required a better understanding of planetary climate (Masco, 2010). This new notion of a planet under threat generated unprecedented levels of investment in scientific infrastructure and personnel. It offered a profitable opportunity for researchers "who were clever enough to offer scientific endeavours and goals aligned with the Cold War obsession with science that focused strongly on the physical and Earth sciences" (Heymann & Dahan-Dalmedico, 2019: 3). The **International Geophysical Year** of 1957–1958 offered a new scientific vision of planetary research and understanding. The international cooperation of scientists offered only a thin veil over what was in fact a rapidly increasing political and military rivalry for scientific prowess.

Out of these Cold War conditions emerged new scientific monitoring technologies, such as Earth-orbiting satellites, instrumented aircraft, radiosondes, and deep-sea instruments. These technologies vastly extended the volume of weather and climate data available. They rendered what had largely been two-dimensional representations of climate into three-dimensional ones, now stretching to ocean depths and atmospheric heights. New analytical and simulation tools were also forged, such as isotope analysis and electronic computers. With respect to climate these investments were directed most profitably towards better understanding the global carbon cycle – especially oceanic and biological processes – and improved atmospheric modelling (Hart & Victor, 1993). The interest in the carbon cycle was fuelled by the questions – now of growing significance – whether and why atmospheric concentrations of carbon dioxide were increasing. Knowing the relative size of planetary sources and sinks for carbon dioxide was essential to this task, as pointed out by Roger Revelle and Hans Suess (1957). The parallel rise of atmospheric modelling was inspired by the scientific instinct to predict and, potentially, control the world's climate (von Neumann, 1955). The development of electronic computation – a product of the Second World War – helped make possible the solving of multiple nonlinear differential equations describing the atmosphere's behaviour.

The development of computer models of global climate during the 1960s fuelled enthusiasm for ambitious political projects. These included practical applications of great value – such as numerical weather prediction – but also more speculative goals concerning weather and climate modification. These latter held strong appeal in an era of geopolitical ambitions and military rivalry. They applied equally to the Soviet Union as to the USA. For example, in the early 1960s a semi-popular book was published in the Soviet Union titled *Methods of Climate Control* written by Nicolai Rusin and Lila Flit (1962). It dwelt on the potential for harnessing advances in science and technology to exploit nature in ways beneficial to the Soviet economy (Oldfield, 2013). These Russian authors proposed two new targets of technological intervention related to climate. One was the modification of regional climates through large-scale drainage, irrigation, and afforestation schemes. The other was altering local weather through influencing atmospheric processes such as cloud formation and behaviour. Other Soviet scientific work related to various aspects of climate modification – often with regard to improving agricultural productivity or opening up the Arctic to shipping – including the seeding of clouds, hail suppression, and the diversion of rivers (Lamb, 1971).

The emerging view of the Earth's climate as a single interconnected and predictable system inspired the human imagination in other ways. For example, it gave rise to **the Gaia hypothesis**, first proposed by James Lovelock and Lynn Margulis in the early 1970s. Although viewed as too mystical by some scientists, the idea of Gaia was influential within new environmentally shaped social movements. Gaia offered a provocative metaphor for thinking about the Earth as one, which embraced various disciplines such as climatology, biogeochemistry, geochemistry, evolutionary biology, ecology, and complexity sciences. Lovelock's Gaia conceived of Earth as a "living organism", a system constituted by myriad organisms and the environment with which they interacted. This thinking converged with the new global infrastructures of planetary-scale monitoring and computerised calculation to develop a fully scientific notion of climate. The new research paradigm of "Earth System science" was firmly established in 1986, when NASA – under the chairmanship of Frances Bretherton – issued its influential eponymous report (NRC, 1986). Earth System science offered a panoptic and integrated view of the planet's climate, a view that makes climate amenable to scientific enquiry, to numerical simulation, visualisation, and prediction – and hence presumptive control (Steffen et al., 2020). At the heart of NASA's vision was a fully systematised and quantitative understanding of climate, captured in the famous Bretherton diagram (Figure 2.4), and the goal of predicting future climate through advanced computer modelling.

As happened during the decades of European high imperialism, the politics of the Cold War powerfully shaped the antecedents of today's scientific understanding of climate change. The military attachments of this new science – together with the ever-present threat of nuclear annihilation – also laid the ground for climate change becoming seen by the late 1980s as a global security threat. This framing was given an early outing in the early 1980s when a group of American scientists led by Carl Sagan (1934–1996) formulated the "nuclear winter" hypothesis. Nuclear winter – the severe global cooling that would likely follow a massive exchange of thermonuclear warheads between the USA and Soviet Union – linked military, nuclear, and climate concerns in a symbolically powerful way. The salience of Sagan's and others' work on this prospect offered the first widespread public recognition – especially in the USA and western Europe – of the cultural power of global climate models as simulation tools of the future.

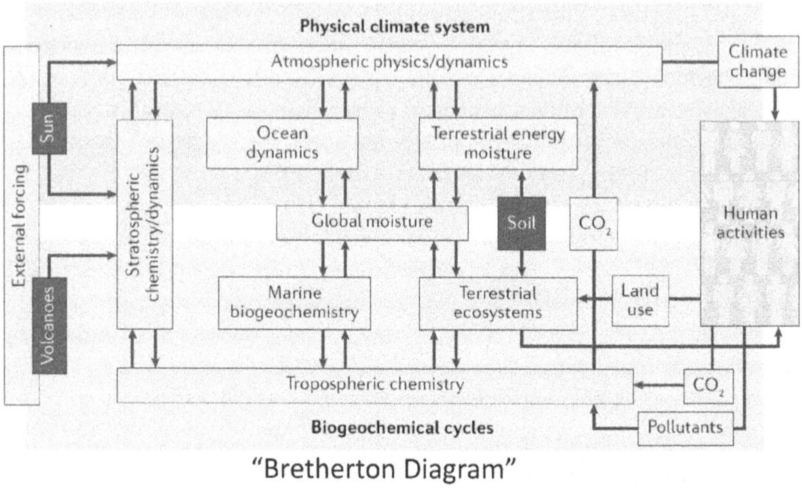

Figure 2.4 The original 1986 NASA/Bretherton diagram depicting the Earth System and its interactions.
(*Source*: Steffen et al., 2020).

United Nations patronage

During the 1980s, a coalition of European and American climate scientists and Western environmental NGOs began to position anthropogenic climate change as an environmental threat on par with thermonuclear war. A series of three retreats held in the Austrian mountain town of Villach in 1980, 1983, and 1985 resulted in increasingly forceful pronouncements about the prospect of significant climate change. By the late 1980s, as one global security threat receded – that of a catastrophic exchange of nuclear weapons – a new one was gaining ground: The fear of global-scale climatic change induced by human activities. The fear of anthropogenic climate change filled the psycho-political space vacated by nuclear war. Over the previous decade, the first modelled predictions of future climate change had combined with the campaigning urgency of environmental groups such as Greenpeace and the Sierra Club. National governments in the Western democracies began to take notice of this modernist reincarnation of the older historical-cultural idea of climatic change (Chapter 1). But climate change was now neither the result of divine command nor a parochial consequence of local human transformation of the land. Climate change now constituted a "change in the global weather" attributable to the fossil-fuelled energy economies and consumption cultures of industrialism (Ross, 1991).

A signal moment in this "securitisation" of climate change was the international conference convened in Toronto in June 1988 "The Changing Atmosphere: Implications for Global Security". The Toronto Conference – held well after *glasnost* ("openness") was underway in the declining Soviet Union and just 18 months before "the Wall" came down in Berlin – was a gathering of a rather eclectic group of political leaders, scientists, and NGOs. One can see in the Conference Statement the transfer of existential anxiety from thermonuclear war to climate change. Climate change represented a "major threat to international security . . . second only to a global nuclear war" declared the statement, urging world leaders that it was "imperative to act now" (Allan, 2017). It was not until 2007 that climate change was formally debated by the UN Security Council (see Chapter 9). But the Toronto Conference catalysed the creation by the United Nations of the two central transnational institutions of global climate governance that survive today: The IPCC in 1988 and the UN Framework Convention on Climate Change (UNFCCC) in 1992.

From the late 1980s onwards, new networks of international scientific collaboration began to form, fuelled further in the 1990s by claims of a new world order and "the end of history". The new internationalism of the immediate post–Cold War decade proved with hindsight to be a naïve illusion. But the IPCC rapidly established its pre-eminence in orchestrating the new global science of climate change. Climate scientists found that the watchful patronage of the American and Soviet military establishments was replaced by the increasing orchestration of global climate science by its new patron, the United Nations. Although defending its mandate to assess science and not to direct policy – "producing policy relevant, but policy neutral knowledge" – the IPCC cannot really offer a "view from nowhere" (cf. Haraway, 1988; Shapin, 1998). Climate science may no longer be influenced by the political view from Washington or Moscow, but the IPCC assessments nevertheless offer a view of climate change heavily shaped by the centres of scientific power located, initially, in North America, Europe, and Japan. Its knowledge remains geographically located, even if its locational anchors are slightly more dispersed.

As we have seen earlier, scientific conceptions of climate change pre-dated the creation of the IPCC. Nevertheless, in the past 30 years the IPCC has powerfully shaped the way in which the idea of climate change has become inscribed in contemporary geopolitics and in the public imagination. The IPCC releases its mammoth Assessment Reports (ARs) every five to eight years – in 1990 (AR1), 1996 (AR2), 2001 (AR3), 2007 (AR4), 2013/2014 (AR5), and 2021/2022 (AR6). Through these highly staged interventions, the IPCC exerts significant power over the practices of climate science – for example through epistemic paradigms, public funding, disciplinary formations, and expertise. It also shapes in significant ways the politics of climate change. We look briefly at these effects in turn.

Influencing scientific practice

One of the original rationales for the existence of the IPCC was to make future climate more accessible to scientific investigation and assessment. To do this, the IPCC adopted – and in turn reinforced – an epistemic paradigm rooted in Earth System science and climate models. Under the patronage of the IPCC, a complex assemblage of conceptual, material, and social technologies has been designed and tuned to offer authoritative claims about the climatic future. This assemblage has elevated climate science into the league of Big Science relying upon Big Data, competing for resources and prestige with high energy physics, neuroscience, astronomy, and genomics. At the heart of this paradigm is the technique of computer model simulation. Numerical models of the Earth System have become the essential means through which future climate can be known. The IPCC cultivated the practice of "model intercomparison", which helps establish the credibility and social authority of models (Heymann & Dahan-Delmedico, 2019). This move towards formal intercomparison of models started in the early 1990s with the first Atmospheric Model Intercomparison Project (AMIP1), evolving into the current incarnation which is CMIP6 (the sixth Coupled Model Intercomparison Project). There are now also "MIPs" for related models of the biosphere, crop yields, geoengineering interventions, and integrated assessment.

To develop quantified predictions of the climatic future, Earth System modelling requires descriptions of possible world futures – scenarios – to enable simulated exploration of their implications for climate. The IPCC has therefore exerted considerable influence over the practice of scenario-making, both qualitative and quantitative. The most prominent of these scenarios have been those in the Special Report on Emissions Scenarios (SRES) published in 2000 (and developed for AR3 and AR4), the Representative Concentration Pathways (RCPs) developed in 2010 for AR5, and the Shared Socio-Economic Pathways (SSPs) published in 2018 for use in AR6. The SRES was developed explicitly by the IPCC, whilst the RCPs and SSPs were indirectly commissioned for the IPCC. The application of these scenarios to climate modelling has not been without controversy.[2] Beyond models and scenarios, the IPCC has shaped scientific practice in other ways. For example, it has reified concepts such as **Global Warming Potentials**, **net-zero emissions**, and the social cost of carbon, and promoted the widespread use of **integrated assessment models** (IAMs). These innovations in climate science would have been unlikely without the voracious demand of the IPCC to package knowledge in forms that are accessible to policy-makers.

Consensus-building has been another central motif of IPCC knowledge assessments. But questions arise about what expertise, which experts, and how consensus is reached. As we will see in later chapters, this drive for consensus exposes the IPCC to criticisms from civic actors about its knowledge formation processes. It attracts claims that the IPCC is either excessively conservative or else unduly alarmist. The drive for consensus can lead to the IPCC issuing the vaguest and blandest of statements or, in the eyes of some, "erring on the side of least drama" (see Chapter 5). The drive for consensus can also exclude minority dissenting viewpoints within climate science or marginalise important questions that may be deemed too value-laden for the idea of consensus to be a sensible means for capturing diversity and disagreement.

The IPCC has also established a de facto set of expectations and definitions about *what* counts as "climate expertise". This therefore determines *who* counts as a climate expert and where such expertise is to be found. A relatively small group of authors and institutions significantly shape the IPCC reports (Hughes & Paterson, 2017). Experts drawn from the natural sciences and from economics are more favoured than those from the social sciences or humanities. Institutional biases are also evident. For example, a study by Corbera et al. (2016) showed persistent North–South inequalities in the authorship of the Working Group 3 (WG3) report in AR5, dealing with mitigation options. A small number of selected IPCC experts regularly co-author papers and a very select number of, mostly, US and UK institutions operated as training sites for IPCC WG3 authors (Figure 2.5). This study revealed a narrowly drawn epistemic community exerting disproportionate influence over the IPCC's assessment of climate mitigation. Biases are also evident in terms of the profile of the enlisted experts. For example, Gay-Antaki and Liverman (2018) show that since the IPCC formed the percent of female lead authors has increased only slowly, from 5 per cent in AR1 (in 1990) to 25 per cent in AR5 (2014). Many of the 100 female IPCC authors Gay-Antaki and Liverman surveyed expressed the view that women were not just poorly represented in the IPCC. They were also poorly heard. And that they encountered barriers to participation beyond their gender, including ethnicity, nationality, command of English, and discipline.

Shaping climate policy

The IPCC does not merely influence and use science to predict future climate. It also alters the climate it is seeking to predict by tacitly influencing policy. The IPCC has long self-policed its mantra of offering "policy relevant, but policy neutral" scientific evidence to governments. But we have seen earlier how the knowledge it puts into circulation – its development of scenarios and use of concepts like carbon budgets and net-zero emissions – cannot but influence the policy process, whether advertently or not. In 2015, the incoming chairman of the IPCC, the South Korean Hoesung Lee, explicitly called for future

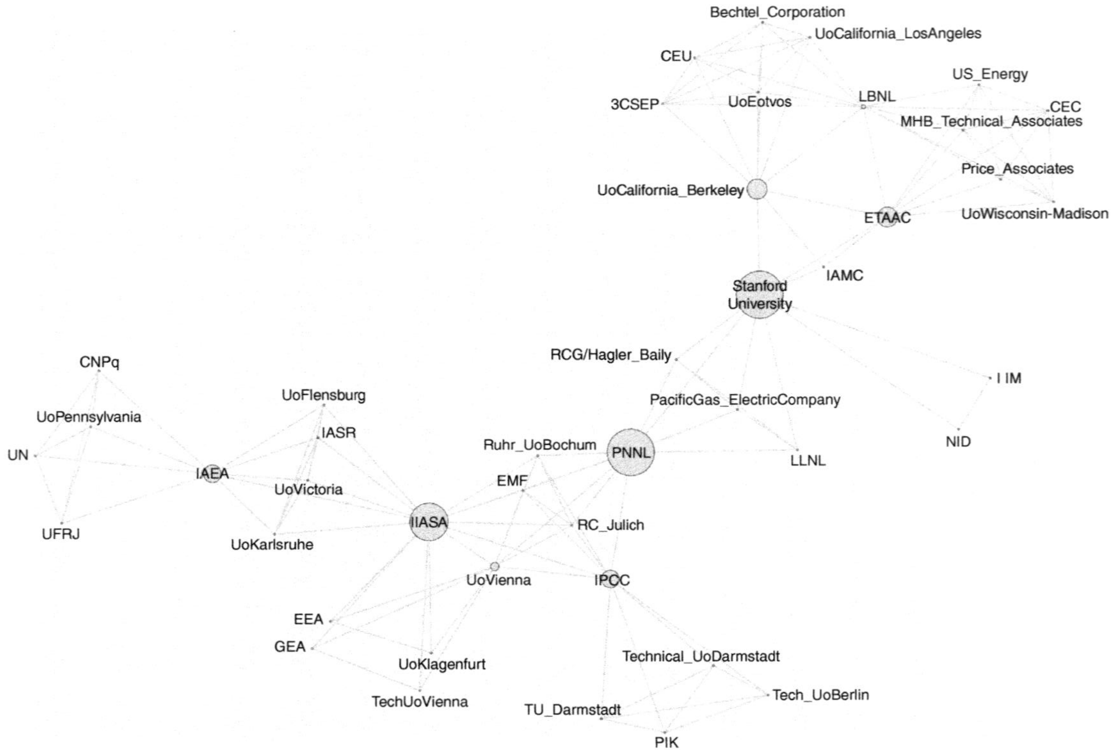

Figure 2.5 Institutional linkages of the top 20 authors in the co-authoring network of IPCC AR5 WG3 lead authors. The size of the node indicates the strength of the interconnectedness between that institution and the others. Within this IPCC author network, the dominance of three institutions is clear – IIASA (Austria), PNNL (USA), and Stanford University (USA).

(*Source*: Corbera et al., 2016).

IPCC assessments to focus less on the state of knowledge and more on offering and evaluating solutions.

Institutions like the IPCC require the rhetorical boundary between science and politics to be constantly reiterated. Yet in practice the location of this boundary is constantly being redrawn. This redrawing enlists "the participation of actors from both sides of the boundary" and often involves the creation of what science and technology studies scholars call "boundary objects" – tools or items of knowledge that are flexibly employed in different social worlds but for different purposes (Beck & Mahony, 2018: 2). Temperature targets like 2°C or 1.5°C would be one example of boundary objects that straddle the worlds of science and politics and that therefore structure action in both (see Box 9.1). The IPCC has also influenced the politics of climate change – and therefore the future course of climate itself – through its endorsement and reliance on IAMs. These models calculate future development pathways that seemingly hold open the possibility of limiting global warming to 1.5°C, but only through the large-scale deployment of carbon dioxide removal technologies (see Chapter 9). These imaginative futures fold back into the international politics of climate change, influencing both the rhetoric and practice of national policy-makers.

Climate nationalisms

It is undeniable that the IPCC has played the central role for over three decades in bringing scientific knowledge about climate change to bear on politics. The IPCC is a creation of the UN and is governed by the UN. The world's governments have therefore, for the most part, paid attention to the IPCC's statements about why the world's climate is changing and how it may change in the future, whether or not they welcome the statements. And successive IPCC reports have been adopted as offering an "anchor" of reliable knowledge for many different political actors beyond the UNFCCC – for example, development agencies, businesses, financial institutions, climate advocacy groups. Yet the legitimating power that adheres to the IPCC with regard to climatic knowledge is in another sense anomalous. Intellectual historians have shown how possessing scientific prowess and calculation have often been viewed as integral to the formation of the modern nation state. And "yet the transnational spaces of knowledge generation and political action associated with climate change seem to challenge territorial modes of political order" (Mahony, 2014: 109). Some nations have perceived epistemic authority seeping away from the nation state and towards the transnational authority of the IPCC. And this has been resisted. Some nations have challenged the IPCC's authority to pass decisive judgement about what is or is not known about climate change within their territorial jurisdiction.

It led to expressions of what might be called epistemic "climate nationalisms". One example of this is how countries have appropriated the development of projections of future climate as a largely national activity. Rather than rely on the generic global and regional projections of future climate issued by the IPCC, many nations have developed their own "national climate scenarios" (Skelton et al., 2019). Similarly, countries such as Brazil, China, Japan, the UK, and the USA, with their more developed scientific capacity, have made moves to establish their own national climate assessment capability. If not supplanting the assessments of the IPCC, these moves certainly reclaim a degree of national ownership of projected climate futures. The development and operation of global climate models also displays similar traits of climate nationalism (see Box 2.2 for the case of the UK's Hadley Centre model).

A more egregious example of the tension between the transnational and national ownership of climate knowledge erupted in India in 2010. This controversy was rooted in long-standing neo-colonial antagonism on the part of India over scientific claims emanating from what is perceived as Western-dominated science. The controversy concerned poorly founded statements in the IPCC's AR4 report about the future of India's climate and glaciers. In response, the Indian Government established an Indian National Climate Change Assessment to reassert territorial ownership over its future climate (Mahony, 2014). Similar tensions emerged in the Netherlands around the same time, triggered by "errors" found in the IPCC's AR4 report about the vulnerability of the Dutch population to rising sea level.

BOX 2.2 THE HADLEY CENTRE CLIMATE MODEL

Global climate models have come to play a central role in both the science and politics of climate change. Climate models are computerised simulation tools that seek to reproduce the behaviour of a comprehensive Earth System, tools that are increasingly relied on for making climate predictions. There are today over 30 climate modelling centres, operating in 17 different countries, each with their own (more or less) bespoke model. The results from these models form the cornerstone of the IPCC's climate assessments. Yet these models do not exist in some placeless, global realm, beyond the reach or influence of national climate politics. Although used by the IPCC in its global knowledge assessments, each individual model is shaped by the history and geography of its own national political culture. The influence of national patronage in the funding, design, promotion, and epistemic status of climate models is unavoidable, even if the mathematical codification of basic physics remains portable. Geopolitics intrudes into climate science. Not all climate models are created equal; not all models carry the same weight in some virtual "climate model democracy" (Knutti, 2010).

The national characteristics of one such model have been explicated in a study of the origins of the UK's Hadley Centre climate model (Mahony & Hulme, 2016). This model was formally "adopted" – and generously funded – by the British government in May 1990 under Conservative Prime Minister Margaret Thatcher. This reconfigured relationships between national climatological expertise and political authority. Climate scientists were brought into a closer relationship with government policy-makers. The motivation on the part of the British government to fund this enterprise was driven by a desire for an independent "national capability" for climate prediction, one that would sever the state's reliance on either American or European models. The nationalist impulse was to develop "a trusty model of one's own" (ibid: 465).

There was therefore something "very British" (ibid) about the way the Hadley Centre and its climate model came into being. A British civic epistemology of pragmatic empiricism and independent judgement informed and guided the political decision by the British government to underwrite this model. It was funded not only because of scientific ambition to strengthen British scientific prowess and prestige. It was also founded through national political instinct. In her speech at the opening of the Hadley Centre, Thatcher emphasised the need for investment in modelling to "ensure that what we do is founded on good science" and "to estimate the detailed distribution of the effects of global climate change". Civil servants sought to fund a British climate prediction capability that would appeal to such sensibilities. The founding of the Hadley Centre by the British government in 1990 illustrates how nation states are keen to exercise and strengthen their own epistemic sovereignty with respect to climate change. It is a process observable in other national jurisdictions, such as Australia, Brazil, China, Germany, and India.

(Continued)

Read:

Mahony, M. and Hulme, M. (2016) Modelling and the nation: institutionalizing climate prediction in the UK, 1988–92. *Minerva*. 54: 445–470; See also: Howe, J.P. (2014) *Behind the Curve: Science and the Politics of Global Warming*. Seattle: University of Washington Press.

But resistance to the international knowledge-forming protocols and claims of the IPCC can also emerge from beyond the state. Climate advocacy groups positioned differently along the ideological spectrum push back against the monopolistic epistemic authority of the IPCC. For example, new social movements like Extinction Rebellion (see Box 5.3) amplify claims about the inappropriate conservatism or scientific reticence of the IPCC. This lies behind their call upon governments to "tell the truth" about climate change, implying that the IPCC's assessments do not adequately do so. Conversely, think-tanks such as the libertarian Heartland Institute in Chicago USA, accuse the scientists enrolled by the IPCC – and therefore the institution itself – as being hopelessly compromised by liberal or environmental ideologies. This prompted the Heartland Institute to darkly mimic the IPCC by publishing a series of reports during the 2010s issued by the "Nongovernmental International Panel on Climate Change" (the NIPCC). We will examine in more detail in Chapter 5 and Chapter 4, respectively, these contrasting forms of civic resistance to the IPCC.

What these various examples reveal is that there is no single uncontested scientific knowledge about climate change, whether issued from the IPCC or from within national jurisdictions or indeed by other civic actors. The knowledge assessments of the IPCC, authoritative within certain transnational institutions and settings, become the battleground upon which struggles are conducted to claim "sound science" as a legitimation of different policy prescriptions. Authoritative scientific claims about the future of climate become a powerful legitimising resource for which different political interests will compete. This points to the intensely political nature of climate science and the tensions that exist among international, national, or local and subaltern (see Chapter 6) ways of knowing climate.

Chapter summary

This chapter has shown that there are distinctive geographies and histories to the science of climate change. Geography matters to the conduct of science. People, places, patronage, and politics have interacted through particular sites of scientific knowledge production – glaciers, satellites, research ships, laboratories, computers, conference rooms, and websites – to establish climate change as a phenomenon knowable *to* science and hence predictable *by* science. But history matters too. To understand the achievements of climate science in giving a scientific explanation of the physical manifestation of climate change, an appreciation of the world's political history over the past two centuries is necessary. Scientific knowledge about climate change has been profoundly enabled and shaped by the exercise of political power: By the rise of European Empires, by the geopolitics of the Cold War, and, in recent decades, by new international institutions of the United Nations.

This observation illustrates a broader point. Scientific knowledge can never fully escape the cultural values, historical contingencies, and political contexts through which it is made. This is true even though such knowledge *appears* universally authoritative when seeming to travel easily across national borders. Science in general – and climate science no less – always struggles to justify and deliver its promise of transcendent timeless and placeless truth. "Knowledge, of whatever sort, can never arise independently of culture, from human ways of doing things. And so knowledge of climate always carries with it beliefs and values about the world it is seeking to describe" (Mahony & Hulme, 2018: 410). This chapter has shown, for example, how distinct national styles of institutional science continue to direct and guide the design and application of global climate models. These models simulate *global* climate. But they reside in – and are shared by – different *national* political cultures, with the inevitable subjectivities around cognitive styles and social norms that this entails.

The critical perspective on the science of climate change developed in this chapter emphasises that all scientific knowledge about climate change is not just shaped by history and geography. It is also shaped by political power. Three important implications follow from this insight, which will become visible in later chapters of the book. First, the science of climate change is inseparable from the wider political cultures within which such science arises. Climatic knowledge is profoundly shaped by imperial, superpower, international, and national interests and mandates. Second, scientific accounts of climate change marginalise subaltern knowledges. Climate science leaves out non-scientific ways of knowing climate change that are profoundly rooted in the cultural histories of specific places. And, third, the previous two realities inevitably create political tensions between abstract "global kinds of climatic knowledge" and more human-scale, intuitive, and situated understandings of climate change.

Notes

1 The History of the IMO. https://public.wmo.int/en/about-us/who-we-are/history-IMO [accessed 15 July 2020].

2 For example, the most extreme of the RCP scenarios, so-called RCP8.5, has been very widely used in impact and policy analysis as though it were a business-as-usual scenario. In fact, RCP8.5 projects such a large increase in CO_2 emissions this century that it offers an increasingly implausible view of the future and yet it continues frequently to be used to depict the world's business-as-usual future (Hausfather & Peters, 2020).

Further Reading

There are several excellent books that offer a critical perspective on how the scientific understanding of global climate change developed. An early classic is **Jim Fleming's** *Historical Perspectives on Climate Change* (2nd edn.) (Oxford University Press, 2005), which progresses the story from European Antiquity to the immediate postwar period. **Deborah Coen's** account of climate science in the Habsburg Empire of the nineteenth century, told in *Climate in Motion: Science, Empire and the Problem of Scale* (University of Chicago Press, 2018), is one of the best for understanding the importance of imperial networks of power and patronage in the making of climatic knowledge. A recent addition in a similar vein but with more global reach is **Martin Mahony** and **Sam Randalls'** edited collection *Weather, Climate and the Geographical Imagination: Placing Atmospheric Knowledges* (University of Pittsburgh Press, 2020). **Coen's** entry *The Advent of Climate Science* (January 2020) in the online Oxford Research Encyclopedia (https://oxfordre.com/climatescience) is a useful summary of climate science up to the First World War. For understanding the climate science of the past 70 years, in particular the role of models in this enterprise, **Paul Edwards'** *A Vast Machine: Computer Models, Climate Data and the Politics of Global Warming* (MIT Press, 2010) is the place to start, while **Josh Howe's** *Behind the Curve: Science and the Politics of Global Warming* (University of Washington Press, 2014) does the same for the politics of climate science, although from a narrowly American perspective. For a personalised and entertaining version of the story told by Howe, read **Steve Schneider's** *Science as a Contact Sport: Inside the Battle to Save Earth's Climate* (National Geographic Society, 2009). In *The Warming Papers: The Scientific Foundation for the Climate Change Forecast* (Wiley-Blackwell, 2011) climate scientists **David Archer** and **Ray Pierrehumbert** offer a useful compendium of 18 classic scientific papers, starting with Jean-Baptiste Fourier in 1824 and ending with Jim Hansen in 2005. Two books to be recommended for understanding science and scientific knowledge more generally as inherently social and situated are **Bruno Latour's** *Science in Action: How to Follow Scientists and Engineers Through Society* (Harvard University Press, 1987) and **David Livingstone's** *Putting Science in its Place: Geographies of Scientific Knowledge* (University of Chicago Press, 2003). And for an excellent introduction to the importance of the Cold War for science in general read **Audra Wolfe's** *Freedom's Laboratory: The Cold War Struggle for the Soul of Science* (John Hopkins University Press, 2019).

QUESTIONS FOR CLASS DISCUSSION OR ASSESSMENT

Q2.1: "The foundations of today's science of global climate were established by nineteenth-century European imperialism". To what extent do you agree with this statement?

Q2.2: Evaluate the claim that the Cold War was essential for establishing the reality of anthropogenic climate change.

Q2.3: To what extent has the IPCC excluded non-scientific ways of knowing climate change? Does this matter?

Q2.4: Are changes in climate better understood scientifically as global or as local phenomena? For what purposes might this be so?

Q2.5: Choose one the following metaphors that have been used to communicate the idea of climate change – "Gaia", "tipping points", "the greenhouse effect", "planetary boundaries". Critically assess the consequences of its usage for climate science and for public policy.

Section 2

FINDING THE MEANINGS OF CLIMATE CHANGE

How do people make sense of a changing climate? How do people make sense of something that on the one hand is both physically real and discursively unavoidable in the contemporary world, but that – at the same time – exceeds human senses and the imagination? Earth System scientists and literary critics alike grasp at the intangibility of climate change. Historian of science Paul Edwards titled his 2010 book about the history and politics of climate modelling *A Vast Machine*, an attempt to capture something of the ambition, scale, and power of Earth System science to simulate the climate system in a computerised machine. This reflects the scientific impulse to explain the complexity of the world through simulation: If reality can be simulated, it can be explained. From a different vantage point however, eco-critic Timothy Morton has described climate change as a "hyperobject" (Morton, 2013). This is something that transcends the ability of any one person to comprehend its scale, reach, and significance. Whether one approaches climate change as a scientist, a literary critic, or as a regular citizen, there remains the common problem. Beyond its mere "facticity", what does the idea of climate change *mean* for different people?

The relationship between facts and meanings is what Candis Callison alludes to in *How Climate Change Comes To Matter* (2014), when she talks about "the communal life of facts". The impersonal scientific facts of climate change – one sort of truth – fail to engage the whole person, however clearly these facts may be communicated. Facts alone fail lamentably to establish another form of truth: An account of what these facts *mean* in human and cultural terms. Without this latter notion of truth, without establishing the social meaning of facts, there can be no participation in communal life, no inspiration, no direction, no hope. For Callison, establishing "truthful meanings" of climate change is crucial. Such meanings matter for

> adjudicating debates, deciding what side one wants to be associated with and 'what's at stake' [with climate change], and

> knowing who and what can be trusted. Not only science, but *the social* determines ethical and moral value – and the consequent rational to act – helping to resolve [climate] challenges to long-held ideals and norms.
>
> (Callison, 2014: 244; emphasis added)

The chapters in Section 2 point towards these "second truths" of climate change. They offer different answers to the question "what does climate change mean?" In these six chapters I examine how different meanings of climate change are constructed, sustained, and mobilised in contemporary public life and political discourse. How people come to know and to understand climate change is intrinsically bound up with the forms of social, moral, and political order they explicitly or implicitly espouse. How people act in response to climate change reveals how they think about the future and the human possibilities of shaping that future.

This approach to explaining the idea of climate change recognises the central importance of "passing judgement on the facts", to paraphrase philosopher Hannah Arendt. It stems from the conviction that humans are conscious, reasoning, and morally judging beings. We are not mere automata who react unreflexively to physical stimuli or to communicated facts about physical stimuli. Human actions are guided – as I believe they should be – by *interpretations* of scientific facts, by the exercise of judgement. Actions are not determined by the facts in themselves. People act in the world according to the meanings they construct, meanings that make sense of their physical existence and personal subjectivities. Put differently, if people are to act politically and purposefully in the world, they need to know what facts mean. This is as true for collective policy making as it is for individual behaviours. Facts only gain significance once they become enlisted in some larger narrative or interpretative scheme (Daston, 1991). Thus, the mere *fact* that physical climates are changing – or that cumulative human actions collectively are re-shaping the world's climate – does not *in itself* demand any specific or self-evident response. Even less does it demand a universal one.

Each of the six chapters in this section describes a broad interpretative position on how and why climate change matters for different social formations and collectives, for different people in different places. Or to put this differently, each chapter examines how and why climate change becomes a matter of truthful concern beyond merely a matter of truthful fact. The first three chapters in this interpretative task are grouped together under the description of "*science-based*" positions, the following three chapters as "*more-than-science*" approaches. Let me elaborate.

"Science-based" positions

The respective positions on climate change described in Chapters 3, 4, and 5 are those of reformism, contrarianism, and radicalism. They reflect contrasting combinations of value commitments and narrative convictions by which the facts of climate science are to be judged and inspire very different responses to climate change. Yet all three positions share the modernist view that science offers cumulative, reliable, and culturally authoritative forms of knowledge. In this view, science is believed to offer the possibility of foundational and relevant knowledge about climate and its future dynamics, even if some of the specific claims of individual scientists or scientific institutions might be contested. As we saw in Chapter 2, this trust in science is epitomised by the creation – and subsequent valorisation – of the IPCC as an authoritative institution to assess scientific knowledge about climate and its changes and to make this knowledge relevant for policy making. For each of these three positions on climate change, "getting the science right" is a prerequisite for acting politically about climate change.

Or at least this is what is claimed rhetorically. In each of the positions described, science is conceived as a self-validating system of knowledge that – with respect to climate change – offers a powerful means to inspire and guide human action in the world. They may diverge about the specifics of whether, when, and how scientists should reach consensus about climate change. In other words, "reformers", "contrarians", and "radicals" may differ about what exactly *is* climatic truth as revealed by science. But each of the three interpretative positions outlined in these chapters affirms the potential for science – and hence of scientists – to "speak truth to power". The differences among these positions cannot therefore be characterised primarily by the way they understand the relationship between science and politics. In this they broadly concur. Science guides policy; it leads, shapes, informs, and underpins it. Without trusted science, policy would be directionless and adrift.

No, the differences between reformers, contrarians, and radicals revolve around two other considerations. First, how they view the *adequacy* of the current state of scientific knowledge about climate and, second, their stance with regard to the *desirability* of the configuration of the world's (largely) capitalist political economy. Let me explain.

The position outlined in Chapter 3: *Reformed Modernism* rests on twin commitments. A commitment, first, to the reliability and adequacy of scientific knowledge about climate change and, second, a commitment to the belief that the political institutions of modernity – if reformed – can be used to deliver the changes to the social and economic organisation of the world that will arrest climate change. By contrast, the narrative position summarised in Chapter 4: *Sceptical Contrarianism* questions, if not directly challenges, the reliability of the current science of climate change, whilst holding firm to the conviction that the existing economic world order should continue more or less intact. The dangers of climate change can be dealt with around the margins of public life. Chapter 5 offers a third contrasting conviction narrative: *Transformative Radicalism*. This interpretation of climate change rests on the empowering urgency that scientific claims about future climate change are deemed to demand. This is broadly consistent with the reformed modernity of Chapter 3, even if climate radicals are critical of some climate scientists for their undue reticence when speaking in public. In contrast to the two previous positions however, transformative radicalism demands a profound restructuring of the world's political economy and a thorough-going transformation of social and cultural norms if climate change is to be arrested.

As explained in each chapter, there are some significant variations within these three broad positions. They also find different expressions in different world regions. Geography most definitely matters, as will be seen. Nevertheless, this tripartite structure of analysis offers one way of exposing the differences in meanings attached by different political and social formations to the idea of climate change. Since each of these three interpretative positions broadly affirms the cultural authority of science, all three stances are deeply interested in evaluating the state of climate science. In particular, they scrutinise the credibility of the climatic futures offered by the IPCC. To simplify (only) a little, climate contrarians believe the IPCC is compromised by liberal and environmental ideologies and produces alarmist science, whilst climate radicals believe the IPCC is too conservative and errs on the side of overcaution. Climate reformers place the greatest trust in the scientific pronouncements of the IPCC. As in the children's tale of Goldilocks and the Three Bears, reformers believe the IPCC's knowledge claims are "just right", neither too alarmist nor too conservative.

"More-than-science" positions

The second set of three chapters in this section – Chapter 6, 7, and 8 – are grouped together as representing what I call "more-than-science" approaches to establishing the meaning of climate change. In each case, what climate science can reveal about the past causes and future prospects of the world's changing climate is given some form of recognition. But this recognition is granted only on terms that are set by ways of being, knowing, and acting in the world that are drawn from beyond the scientific paradigm. There is no instinctive deference to science as offering the most authoritative way of knowing the material world or of explaining human subjectivity. For the approaches described in these three chapters, science does not – indeed, it would be claimed, cannot – provide an adequate foundation either for motivating political action or guiding personal behaviour.

These three chapters are grouped together not to set up a false conflict between science and culture. After all, science is one of the great accomplishments of human culture. Rather, they are distinguished from the positions on climate change outlined in Chapters 3, 4, and 5 so as to make clear one thing. For many people the idea of climate change is interpreted through more expansive and holistic understandings of the world – and the place of people within these worlds – than can be offered by science alone. Chapters 6, 7, and 8 show that for many people the meaning of climate change resists capture by a modernist worldview, by Western hegemony, or by a scientific epistemology.

Each of these chapters explores different ways in which mythical, personal, and normative knowledge of the world – how humans come to be, to know, and to act – exceeds the forms of climatic knowledge offered either by the natural of social sciences. The assumption that climate science is necessary for understanding the full significance and implications of climate change – and therefore provides an adequate basis for the development of policy – is challenged in different ways. Instead, environmental knowledges that emerge from lived experience or from the humanities – anthropology, history, philosophy, literature, the arts, religious studies – are foregrounded. Chapter 6: *Subaltern Voices* describes what the idea of climate change signifies for cultures and social formations holding diverse cosmologies, in possession of situated knowledges and that have experienced forms of historical oppression. Chapter 7: *Artistic Creativities* reflects on the multiple and provocative ways creative artists seek to represent climate change and its multiple meanings through different cultural media, such as texts, visual imagery, and performance. Chapter 8: *Religious Engagements* considers the enduring importance of religious belief, practice, and identity for human experience of the world. Specifically, I show how the idea of climate change is interpreted through expansive religious narratives of human purpose, solidarity, duty, and destiny.

As with Chapters 3, 4, and 5, there are variations within each of these three broad interpretative approaches for making sense of climate change. Climate scientists might seek – and indeed may attain – expressions of consensus about the veracity of certain scientific knowledge claims. But these "more-than-science" approaches resist attempts to negotiate or impose a consensus about how stories and myth, artistic representations, or embodied religious practices yield a singular meaning to the idea of climate change. They recognise that, for the most part, the meanings of climate change escape the grasp of any single person or culture. They embrace pluralist understandings of the world. On the other hand, these "more-than-science" approaches each recognise that different cultural expressions of meaning and purpose with respect to climate change offer powerful motivational narratives that guide personal and collective reflection, resistance, and action.

Summary

The broad interpretative positions outlined in each of the following six chapters require the inter-penetration of the descriptive – "this is what is happening to climate" – with the normative – "this is what I should do about it". But the "science-based" positions of Chapters 3, 4, and 5 are more likely to deploy normative claims such as "the science says", "let's get the science straight", "listen to the scientists", or "speaking truth to power". The "more-than-science" approaches of Chapters 6, 7, and 8 are more likely to be distinguished by public normative claims, which do not emphasise science. They are more likely to claim, for example, "justice demands", "we must re-imagine life on Earth" or "this is how we should live".

3

REFORMED MODERNISM

. . . Climate change assimilated . . .

Introduction

The international climate diplomat Christiana Figueres comes from a hugely influential and politically successful Costa Rican family. Her father led a 1948 social democratic revolution against a nationalist government and subsequently served three separate terms as President of the Republic. Her brother likewise served as President of Costa Rica in the mid-1990s. And her mother was a diplomat and member of the country's national assembly. Figueres herself progressed through the nation's diplomatic service before entering the world of international environmental diplomacy in the 1990s. Most significantly, she held the role of Executive Secretary of the UN's Framework Convention on Climate Change (UNFCCC) for six years from 2010 to 2016. During this incumbency she chaired six successive meetings of the Conference of the Parties (COPs) to the Convention – those held in Cancún, Durban, Doha, Warsaw, Lima, and, finally, in December 2015 in Paris. More than any other person, Figueres is given credit for securing the historic Paris Agreement on Climate Change (henceforth PACC, or "the Paris Agreement") that emerged from the COP-21 meeting.

In many senses Christiana Figueres embodies the reformed modernist position on climate change that this chapter outlines. In brief, it runs as follows. Climate change must and *can* be arrested; carbon pricing and innovation *can* accelerate decarbonisation; sustainable economic growth *is* achievable. Climate change presents a supreme challenge to modernity – but one that *can* be met. As with other reformed modernists, Figueres is optimistic that the institutions of modernity can be steered in the needed direction. "This is the decade", she says, "in which . . . we have everything in our power. We have the capital, the technology, the policies. And we have the scientific knowledge to understand that we have to half our emissions by 2030".[1] Her optimism is grounded in an international agreement – the Paris Agreement – not many thought feasible. But over the last few years it has set in motion adjustments to markets, technologies, institutions, investments, and policies around the world. Figueres believes the new form of multilevel transnationalism encapsulated in the PACC gives the world its best hope of defusing the dangers of climate change.

The reformed modernist position on climate change that Figueres and others espouse rests on twin commitments. Taken together, these commitments contrast the climate *moderniser* to both the climate *contrarian* and the climate *radical*, whose responses to the idea of climate change are outlined in the following two chapters. The first commitment is to the primacy, reliability, and adequacy of scientific knowledge for guiding global political action. As Figueres says with respect to the IPCC's 2018 Report on 1.5°C, "It is a robust and outstanding example of international cooperation, and an extraordinary source of shared intelligence. . . . Now we all have the information we need . . . to move forward together".[2] The second commitment is to a belief that the political institutions of modernity – if reformed – can deliver the changes to the world economy that will arrest climate change. These characteristics of modernity include, among other things, a globalised arena for science, technology, and innovation; the instruments of a modified (i.e., an environmental) economics; the pursuit of sustainable growth and development guided by rational scientific knowledge; and a commitment to the international political arena afforded by the United Nations. As this chapter will show, reformed modernism embraces a number of different framings, political strategies, and climate policies. But its unifying conviction is that climate change is an urgent and pressing problem to be solved – and that it is one that *can* be solved by reforming the institutions of modernity rather than by overthrowing them. Inheriting her family's social democrat legacy and shaped by her own firm commitment to internationalism, Figueres is no revolutionary.

The lineage of reformed modernism can be traced back to the origins of modernity in Western Enlightenment rationality. This Enlightenment set in motion the ideologies of material and social progress, of democracy and human emancipation, of national self-determination, of a world community of nations and – more recently – of sustainable development. Reformed modernists are optimistic about the future, or at least they are optimistic that the future can be made better. With respect to climate change, these ideological commitments culminated in the adoption by the world's nations of the UNFCCC in 1992. It was not insignificant that this accomplishment followed shortly after the collapse of Soviet-style imperial communism, a collapse that liberal democracy claimed as a victory for Western values.

***

The chapter starts with a brief overview of the main characteristics of reformed modernism inherited from these developments, as applied to climate change.

I then explore this position on climate change using three different lenses: Institutions, economics, and technology. Within each of these domains there are vibrant debates to be had about how modernism should best be reformed. These include the nature of growth and the utility of the social cost of carbon; the relative merits and demerits of carbon markets and carbon taxes; the means and methods of mobilising and directing climate finance; and the necessity and nature of technology innovation policies. I sketch some of the broad contours of the respective arguments and points of tension. A concluding subsection then summarises how the reformed modernist position finds expression in two illustrate formulations: That of *ecomodernism* – as promoted by the Breakthrough Institute in California – and the idea of *ecological civilisation* embedded in the new Chinese model of sustainable development. The chapter draws upon the disciplines of *economic* and *political geography*, *innovation studies*, and *environmental economics*.

Reformed modernism

A modernist reading of climate change inherits many of the commitments to progressive liberalism and developmentalism demonstrated during the latter half of the twentieth century. The roots of these commitments lie further back in the Western Enlightenment traditions of scientific rationalism, state administration, political emancipation, and cosmopolitanism. This combination of rationalism, human rights, and democracy informed the diagnosis of the economic, social, and environmental problems of modernity, which were becoming ever clearer during the 1970s. For example, it underpinned the 1980 report of the Brandt Commission – *North-South: A Programme for Survival* (Brandt Commission, 1980) – and the more influential *Report of the World Commission on Environment and Development: Our Common Future*, the so-called Brundtland Report of 1987 (WCED, 1987). And these commitments continue to offer the central tenets by which the reality and challenge of climate change is today interpreted by reformed modernists. It also inherits the optimistic global environmental institutionalism of the late twentieth century. The new wave of global institutions brought into being during this period used the instruments of high modernity (Scott, 1998) to solve various transboundary environmental problems of the 1980s and 1990s – for example acid rain, hazardous waste, and stratospheric ozone depletion (O'Neill, 2009). Within this period of new internationalism, the Rio Earth Summit of 1992 gave birth to the UN's Framework Convention on Climate Change and to the parallel multi-lateral conventions tackling the global environmental challenges of biodiversity and desertification.

For climate modernists, the challenge of climate change is one to be met primarily by reforming the institutions, economics, technologies, and governance arrangements built up over the past century and particularly since the Second World War. And science is to be relied on to provide the rational knowledge needed to guide strategy formation and tactical decision making with respect to these domains. As exemplified by the case of Christiana Figueres, reformed modernists place great faith in the scientific analysis and prognostications of successive IPCC reports. They hold great political value. While scientific uncertainties about the climatic future are recognised, for climate reformers these reports offer a solid enough basis upon which the paradigm of risk assessment and risk management can be operationalised to guide political action on climate change (for example, King et al., 2015). Consideration of the risks of climate change and the means of mitigating them is to be "mainstreamed" – assimilated – into all aspects of modernity, bolstered by new multilateral international agreements.

Reformed modernists are committed to providing the world with energy that is reliable, affordable, and that does not emit carbon dioxide or other greenhouse gases. They are generally optimistic that innovation policy, philanthropic entrepreneurship, carbon pricing, and green finance – and other newly fashioned instruments of modernity – can deliver this goal. They therefore propose "incremental changes to infrastructure and technology" rather than "transformational changes in economic and social structures" (Heikkinen et al., 2019: 91). Governments need to foster new energy technologies through specific innovation policies (Nemet, 2020). And following the path mapped out by the Brundtland Report in the 1980s (WCED, 1987), reformed modernists also view sustainable economic growth as eminently achievable. Current lifestyles for the affluent can broadly be maintained but need to be enhanced for the poor. Economic expansion is desirable and necessary for those deprived of basic human needs. Private and philanthropic investment and public–private partnerships are vital for this task.

Seeing climate change as "the greatest example of market failure the world has ever seen" (Stern, 2006: 1), reformed modernism promotes the idea that intelligent use of market mechanisms is the most effective way of creating a low-carbon economy and preventing dangerous climate change. A price on carbon is seen as capable of launching a "global energy revolution". Reformed modernists call upon a new cadre of environmental economists to overturn neoclassical economic orthodoxy. Drawing upon the idea of **natural capital**, market values are to be assigned to **ecosystem services** to drive forward market reform and consumer behaviour. These services include the benefits of a stable climate, which can be quantified through the social cost of carbon. Through these reforms, climate change is turned into a business opportunity, "one of the greatest wealth-generating opportunities of our generation", according to the entrepreneur Richard Branson.[3] Metaphors such as "carbon footprints", "carbon sinks", and "green growth" are used to reshape public discourse (Shaw & Nerlich, 2015). The central tenets of social market capitalism remain intact.

This is only the barest sketch of the reformed modernist perspective on climate change and it includes numerous points of tension, some of which are explored later. For example, the green new deal, ecomodernism, and ecological civilisation are three different variants introduced later in this chapter of the reformed modernist position.

Reforming institutions

Climate change is framed by reformed modernists as a challenge for global collective governance. This implies the need to strengthen cooperative arrangements between state and nonstate actors, between public and private entities, between formal and informal institutions, and between international, national, and subnational jurisdictions. It is within such enhanced cooperative institutional structures that the actions of groups and individuals will be regulated in ways that will defuse the most severe dangers of climate change. This commitment to reforming institutions in pursuit of progressive goals is one of the hallmarks of modernity.

International institutions

Reformed modernists retain an unyielding belief in the power and necessity of international institutions to cultivate the interstate cooperation that is seen as a prerequisite for any sustained progress towards climate mitigation goals. We saw this in the case of Christiana Figueres. The impulse for expanded cooperation between states and towards institutions of global governance can be traced back over 200 years of modernity (Mazower, 2012). Seminal moments in this history include the formation of institutions for transnational cooperation in science and medicine in the nineteenth century, the Bretton Woods agreements and the inauguration of the United Nations in the 1940s, and the new internationalism of the 1990s.

With regard to climate change, this modernist pursuit of international cooperation is grounded in the 30-year history of the UNFCCC, the most recent manifestation of which is the shared commitment to emissions control embodied in the 2015 Paris Agreement. The institutional architecture associated with the UNFCCC took inspiration from two earlier successes of this modernist impulse. These were the 1970 Treaty on the Non-Proliferation of Nuclear Weapons and the 1985 Vienna Convention for the Protection of the Ozone Layer. But designing the requisite institutional architecture for governing global climate has not been straightforward. Over time, it has become significantly more elaborate and extended. (see Chapter 9). This has prompted political scientists to describe the regulatory activities of different international institutions with regard to climate change – and the relationships between them – as a "regime complex" (Keohane & Victor, 2011).

National institutions

Effective governance in the modern world has been premised on the pre-eminence of the nation state. It is for this reason that the international regimes orchestrated under the United Nations umbrella remain very largely state-centred. For many climate modernists the state therefore retains a central position in the making and shaping of the policy landscapes upon which other actors operate (Giddens, 2009). Yet reform of national state institutions is necessary if the new and difficult challenges associated with climate change are to be met. It is the state that must legitimate, implement, and police regulations, incentives, and policies that will alter corporate, social, and individual behaviours in directions that minimise human impact on climate. Institutional reform at the national scale will always reflect distinctive political cultures. The contrasting cases of a federal democracy (Germany), a constitutional monarchy (the UK), and a central party state (the People's Republic of China) provide suitable illustrations.

By the late 2010s, it was clear that the German energy transition – *Energiewende* – was failing to deliver on the nation's climate mitigation targets. This was largely because of the earlier decision in 2011 to close down all of Germany's nuclear power stations. Coal therefore persisted in the national energy mix, in 2018 still delivering around 25 per cent of the country's primary energy consumption. In 2018 the Federal Government therefore established a "Commission on Growth, Structural Change and Employment" – otherwise known as the Coal Commission – to consider how to transition the country away from its dependency on coal. This state-sponsored commission followed well-established German modes of consensual political process – the so called Enquete Commissions – which involved sustained deliberations amongst all relevant stakeholders (Jasanoff, 2005). The following year the commission presented a comprehensive roadmap for the phase-out of coal-fired power generation in Germany by 2038, which the Federal Government has said it will implement in full.

The UK followed a different path of institutional reform. In 2008, the UK Parliament passed the Climate Change Act, which represented a new institutional arrangement for regulating the relationship between national government and technocratic expertise (see Box 3.1 for further elaboration). For some reformed modernists, the role of the state in responding to climate change suggests a more extensive reformulation of state institutions. This may result in what some commentators advocate as "the green state" (Eckersley, 2004). Such wholesale and centralised reform is more in keeping with the institutional changes seen in China in response to climate change. Over the past decade there has been a growing attention paid to the goal of "ecological civilisation" and in 2018 this ambition was enshrined in the Chinese Communist Party's manifesto (see Box 3.2 later in the chapter).

BOX 3.1 THE UK'S CLIMATE CHANGE COMMITTEE

During the 2000s, there was a growing cross-party desire in the UK to define climate change as a policy issue that transcended the usual four- or five-year lifetime of a single government. Partly as a response to a successful campaign coordinated by environmental NGOs, in November 2008 the UK Parliament passed, almost unanimously, the Climate Change Act. It was one of the first political jurisdictions to institutionalise in legal terms the making and monitoring of national climate policy. The 2008 Act established a Committee on Climate Change – "the Committee" – as an independent, statutory expert body whose primary purpose was to advise the UK Government on national greenhouse gas emissions reduction targets. In establishing this new institution within the constitutional terms of a parliamentary democracy, the act can be seen an example of reformed modernism.

The committee was charged with providing the government of the day with independent advice on setting, meeting, and monitoring carbon budgets. It was also required to prepare the nation for future climate change through a systematic strategic programme of adaptation. In effect the Committee acts as the custodian – but not the executor – of UK climate policy. As with other modernist advisory bodies, it relies centrally on scientific and other forms of technical expertise in guiding its business, not least in crafting a national science-based climate change risk assessment every five years.

The 2008 act brought about institutional *reform* in the UK, rather than radical *transformation* of constitutional decision making with regard to climate change. The act has undoubtedly altered the dynamics of climate policy making in the UK, for example by linking scientific expertise, business, and financial interests with parliamentary committees in a new way. The act requires that government decisions pay careful attention to the advice offered. For example, in June 2019, following the recommendation of the committee, the outgoing Prime Minister – Theresa May – secured parliamentary legislation to strengthen the UK's statutory domestic emissions reduction target to net-zero emissions by

(*Continued*)

2050. On the other hand, the government pressed ahead with its decision to expand London's Heathrow Airport with a third runway, contrary to the recommendations of the committee. The committee did not possess legal power to overturn this decision, but the courts did. Environmental groups and the mayor of London brought a case against the government for its Heathrow decision and in February 2020 the Court of Appeal ruled such a decision "unlawful". This is an interesting example of how the courts become political actors in the governance of climate change (see Chapter 9).

Read:

Lockwood, M. (2013) The political sustainability of climate policy: The case of the UK climate change act. *Global Environmental Change*. 23: 1339–1348; See also: Climate Change Committee (2017) *UK Climate Change Risk Assessment 2017*. London: HMRC; Grantham Research Institute (2018) *10 Years of the UK Climate Change Act*. Policy Brief. LSE: Grantham Research Institute on Climate Change and the Environment.

Non-state institutions

Some climate modernisers place more weight in reforming nonstate rather than state institutions. For them, the focus of reform in response to climate change is on nonstate actors such as businesses, NGOs, regional and municipal authorities, universities, trade unions, and other civic institutions. This might entail the strengthening of corporate social responsibility, implementing local authority green investment schemes, divestment of pension funds from fossil fuel companies, or voluntary carbon offset schemes run by NGOs. This approach to reforming modernity recognises the huge array of institutions across multiple scales, which shape the fabric of contemporary life. This "highly complex institutional environment" (Abbott, 2012: 571) is inevitably implicated both in the changes wrought on society by a changing climate and in the implementation of change. For some climate modernisers this focus on the diverse, polycentric, and fragmented nature of social institutions offers a more plausible way of instigating needed reform (see Chapter 9). Each nonstate institution may on its own have limited reach and impact on the processes influencing future climate. Yet reforming such civic institutions is seen as a path for building political constituencies that may coalesce to yield more powerful calls for reform at state and international levels (Andonova, 2020).

Reforming economics

Economics is concerned with the allocation and management of scarce resources and therefore offers an essential analytical tool for developing strategies and policies for arresting climate change. Three broad schools of economic theory and practice can be recognised: Neoclassical, environmental, and ecological – or heterodox – economics. In its neoclassical form, economics has long exerted a powerful discipline over the institutions of modernity and continues to do so. But reforming modernisers challenge the competence of neoclassical economics for handling the challenge of climate change. Instead, they would advocate for an environmental economics framework. (We will see in Chapter 5 how an ecological economics approach differs.) There are three interrelated pillars upon which the environmental economics of reformed modernism rests: The possibility of securing sustainable economic growth (i.e., green growth); the necessity of appropriately pricing carbon, whether through market mechanisms or carbon taxes; and the need to mobilise and deliver new international climate finance.

Sustainable growth

Reformed modernists believe it is possible to restructure societies and their economies in such a way as to accelerate the decoupling of emissions of greenhouse gases from the market transactions that add value to an economy and that signal economic growth. In other words, sustainable economic growth presumes that a society can reduce its emissions whilst increasing the quantity of goods and services produced and consumed. This is referred to as reducing the carbon intensity of the economy, decoupling economic growth from carbon emissions. To a degree, this has occurred over the past 30 years in the global economy; the annual global decarbonisation rate since 2000 has averaged about 1.6 per cent (Figure 3.1). This decoupling is sometimes referred to as "green growth", which captures the idea of complementarity between environmental protection and economic growth. This idea lies at the heart of a reformed modernist response to climate change. Since the late 2000s, green growth has become increasingly endorsed by some of the largest and most influential organisations in the world (Meckling & Allan, 2020), including the OECD, the UN Development Programme, the World Bank, and the Asian Development Bank.

A seminal intervention that supported this reformist vision was "The Stern Review on the Economics of Climate Change", published by the UK Government in 2006. This was followed shortly after by the idea of a "green new deal" for tackling climate change, a term first used in 2007 by journalist Thomas L Friedman in the *New York Times*.[4] Following the financial crash of 2008, the idea gained greater visibility and, since 2018, a green new deal has been reincarnated in the programmes of centre-left political parties in the USA, UK, and

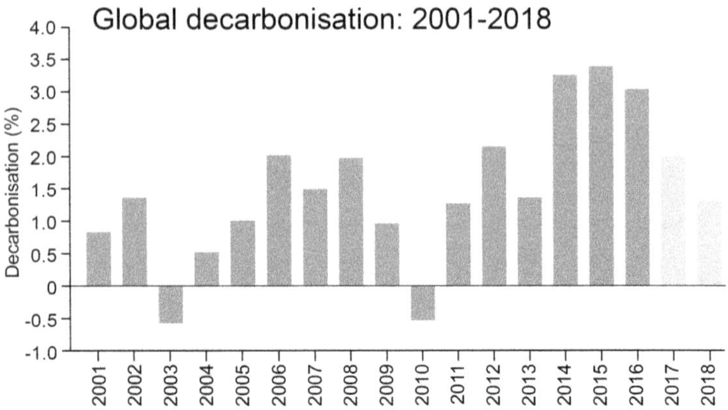

Figure 3.1 Global decarbonisation 2001 to 2018. Annualised percent reduction in the carbon intensity of global economic wealth creation. Negative rates in 2003 and 2010 indicate that in these years the amount of carbon emitted per unit of wealth creation increased (2017 and 2018 numbers are provisional).
(*Source*: IEA and World Bank, courtesy Roger Pielke Jr.).

other nations. (There are different versions of the green new deal and Chapter 5 examines more radical versions.) Achieving sustainable green growth requires improvements in the efficiency of production and manufacturing processes, strong carbon pricing mechanisms, and transitioning energy production to low-carbon technologies (see subsections below). Climate modernists defend this growth goal on the grounds that GDP – the conventional measure of economic growth – has been proven historically to be closely correlated with many indicators of human well-being and development.

Some national economies *have* managed to decouple economic growth from greenhouse gas emissions. For example, in the twenty-first century, France, the UK, the USA, China, and Russia have all decarbonised their economies at rates between 3.7 (the UK) and 2.5 (the USA, China) per cent per year, rates that are about double the world as a whole (see Figure 3.1). Many developing and emergent economies have much lower rates of decarbonisation during this period – India and Indonesia 1.4 per cent, Mexico just 0.7 per cent. For reformed modernists the question remains how fast this decoupling can be accelerated across all nations. The International Energy Agency estimates that to limit global warming to 2°C, the average global rate of decarbonisation required over the next three decades is about 7 per cent per year. No country in the world has ever achieved that rate over a sustained period of time.

More fundamentally, one of the points of tension between proponents and opponents of green growth concerns the compatibility of economic growth with the finite nature of the planet. The processes that produce economic goods and services measured by GDP use raw materials as inputs and generate pollution, carbon emissions, and waste as outputs. The argument hinges on whether, put crudely, unending material growth of the global economy is compatible with the resources – the natural capital – of a finite planet. The more optimistic reformed modernists would be confident – or at least hopeful – that technological innovation can offset what ecological economists such as Eric Neumayer (2007: 297) refer to as the "irreversible and non-substitutable damage to and loss of natural capital" caused by conventional economic growth.

Pricing carbon

For the reformed modernism position, one of the central means for reconciling sustainable growth with arresting climate change is the principle of pricing carbon. The purpose of such pricing is to reduce greenhouse gas emissions in the most cost-effective manner. Environmental economists have argued for several decades that the cost of goods and services should reflect their negative environmental externalities – i.e., the environmental harms that result from the process of production and transaction. Only with such a pricing philosophy in place will markets be compatible with the goals of sustainability. For many reformed modernists, pricing carbon is perhaps the single most important policy intervention that is advocated.

Achieving this objective often splits into two different – but related – discussions. What is the appropriate externality for emissions of greenhouse gases, a quantity usually referred to as the social cost of carbon? And what combination of carbon markets and/or taxes is likely to work most efficiently to provide the best incentive structure to drive down emissions? The social cost of carbon is intended to reflect the social and environmental damage incurred by the incremental change in climatic conditions induced by one unit of emitted carbon. But it is much easier to articulate the principle of the social cost of carbon than it is to calculate a universally agreed value for it. The reasons for this difficulty are twofold, one epistemic and one normative (Pezzey, 2019). The epistemic problem is one of modelling – how can the impacts of future climate change – and their monetary value – be known? The normative problem is one of discounting – how should the needs of people in the future be weighted relative to those living today? Both of these elements are subject to huge uncertainties, which means that estimates of the social cost of carbon range from $5 to $500 per tonne of carbon emitted.

The second point of tension within the reformed modernist position on carbon pricing concerns the question of carbon markets versus carbon taxes. Both market regulation and taxation are well established mechanisms in classical economic theory for the public regulation of transactions, yet the former resonates more with neoliberal instincts and the latter more with **Keynesian economics**. Carbon markets and carbon taxes are often portrayed as alternative policy instruments, but climate reformers can support both instruments. Although the relative merits of either instrument can be exploited in pursuit of climate mitigation goals, the efficacy of either taxes or markets in reducing emissions is in fact poorly known (Haites, 2018). Carbon tax regimes have hardly evolved over 25 years and so there is little experience from which to learn. Most existing carbon tax rates are low relative to levels thought needed to achieve climate change mitigation objectives. In contrast, the design of emissions trading schemes has benefited from different learning experiences in varied market conditions. And yet most carbon markets have faltered in respect of either the efficient allocation of permits or of emissions "leakage" outside the domain of the market (Cullenward & Victor, 2020). Spot prices of carbon in most emissions trading systems are highly volatile (see Figure 3.2), although have generally been too low to incentivise investment and market behaviour in significant ways.

Climate finance

It is not just innovations and adjustments with respect to carbon pricing that climate reformers pursue. There is also the need to mobilise another key institution of modernity in pursuit of climate mitigation goals. This is the regulation of private investment capital and the directing of bilateral and multilateral state development aid. Over the last two decades in particular, the

Figure 3.2 European Carbon Allowance price, euros per tonne of carbon traded. (Reproduced courtesy of *The Financial Times*, July 2020.)

idea of climate change has been used to reform the principles and operation of the postwar development paradigm. This saw modernisation and economic growth as the way to secure human well-being and prosperity. For reforming modernisers these twin principles are now reformulated around flows of investment capital and development aid for leveraging low carbon technologies and climate-resilient development. Thus Article 2 of the 2015 Paris Agreement entails a commitment to "making finance flows consistent with a pathway towards low greenhouse gas emissions and climate-resilient development". This is so-called green or climate financing.

Although an internationally agreed definition of climate finance is elusive, the broad formulation enshrined in the UNFCCC offers an approximation: "local, national or transnational financing – drawn from public, private and alternative sources of financing – that seeks to support mitigation and adaptation actions that will address climate change". Climate financing now operates across multiple scales – multilateral, transnational, state, city, and private sectors. The PACC reinforced earlier commitments from developed nations to mobilise US$100 billion per year by 2020 of new finance for climate action in developing nations. But this target is very far from being realised, with less than a quarter of this amount being committed by 2016 (see Table 3.1). And the scale of investment needed is estimated to be at least an order of magnitude greater than this 2020 target.

If the financial flows of a capitalised world economy are to be redirected to secure the 2°C global temperature target, then up to $1 trillion of new

Table 3.1 The top eight donors and recipients of climate finance averaged over 2015 and 2016. Only public finance flows from "developed" to "developing" nations included. South–South financing – for example from China to countries in the Global South – are excluded.

Largest climate finance donors		*Largest climate finance recipients*	
Country	$m/yr	Country	$m/yr
Japan	10,322	India	2,603
Germany	6,495	Bangladesh	1,357
France	3,671	Vietnam	1,344
United Kingdom	2,618	Philippines	1,296
USA	2,370	Thailand	963
Netherlands	940	Indonesia	952
Sweden	918	Kenya	766
Norway	755	Turkey	665

(*Source*: Carbon Brief, 2018).

investment in low-carbon activities will be needed each year (Bridge et al., 2020). And if the ambition of climate reformers is to extend to securing the resilience of modern infrastructure to future climatic hazards, then the price tag increases dramatically. The OECD estimated in 2017 that $6.5 trillion of new annual investment in climate smart infrastructure – energy, transportation, water, telecommunications, agriculture – will be required globally by 2030. About two-thirds of this will be needed in developing countries. Realising the modernist vision of financing a new low carbon economy will therefore require significant reform to the earlier focus on mobilising *public* capital through development aid and state obligations. Much larger flows of *private* capital will need leveraging to meet this financing challenge, including from nonstate and substate actors such as corporations, cities, and philanthropists. Yet private philanthropic interventions, such as offered in recent years by Jeff Bezos and Bill Gates, have been relatively limited. Bezos committed $10 billion in 2020 to "fight climate change", Gates considerably less than this.

Innovating technology

The *institutions* of modernity might be reformed along the lines suggested earlier. And new *carbon pricing* instruments and financial flows might be introduced to remedy the shortcomings of the market. But a central question still remains for modernists committed to delivering sustainable economic growth and to meeting the expansion of essential energy services for billions of people yet denied them. The world currently obtains around 85 per cent of its primary energy needs from fossil carbon sources. Can this fossil energy be displaced by low- or zero-carbon energy technologies already in existence? And can these technologies deliver the future expansion of energy services *fairly* for all newcomers? The third lens through which to understand the reformed modernist position on climate change is that of technology.

The reformed modernist position is generally techno-optimistic, those who identify as ecomodernists even more so (see later sections). To realise the full benefits of new technologies, climate modernisers call upon notions such as "mission-oriented innovation" or "directed technological change". These calls follow **Schumpeterian** thinking by offering government subsidies for research and development of clean energy technologies. Mariana Mazzucatu's argument in *The Entrepreneurial State: Debunking Public vs. Private Sector Myths* (2013), and in her later book *Mission Economy* (2021), is significant in this regard. Her advocacy for state-led innovation policies is rooted in the conviction that such efforts not only sustain economic growth – a core promise of modernity. Mazzucatu also argues that this is the best route to securing the significant co-benefits of environmental protection and wealth redistribution. The following discussion focuses on three families of technologies for mitigating climate change: Technologies for displacing carbon in the energy system; technologies for capturing carbon from the atmosphere; and technologies for offsetting the global heating induced by carbon. Some of the challenges for climate reformers of each family of technologies are identified.

Generating energy

Reformed modernists set themselves the challenge of satisfying today's and tomorrow's energy needs, whilst at the same time rapidly accelerating the decarbonisation of the energy supply. A central point of contention is whether the current portfolio of renewable energy technologies is capable of delivering these twin goals. All reformed modernists agree that technological innovation is necessary. The differences arise when considering where the focus of innovation should lie. Is the challenge primarily one of scaling-up existing renewable technologies? In which case innovation is about developing and rolling out the new infrastructure necessary to support a massive expansion of renewable technologies. This is the position strongly espoused by advocates such as Mark Jacobsen of Stanford University and his formula of 100 per cent WWS – "water, wind, sun" – energy roadmaps (Jacobson, 2020). Or is the challenge more fundamental, namely the need to innovate across a broader portfolio of zero carbon energy technologies, including **general purpose technologies**? This position, too, is argued strongly by many energy policy scholars (Clack et al., 2017) and by ecomodernists (Symons, 2019). Whichever position is favoured, reformed modernists are likely to be sympathetic Schumpeterians. Directed state policy and public financial support will be essential to deliver the rates of innovation deemed necessary.

Some deeper questions lie hidden in this debate beyond the superficial distinction between renewable-optimists and pessimists (Stephens & Nemet, 2020). One of these questions concerns the perceived relative risks of different energy portfolios. Not least here is nuclear energy, which currently delivers about 5 per cent of the world's primary energy, but around 30 per cent of primary energy from non-fossil sources. Reformed modernists fall on different sides of the question of whether nuclear energy is part of the desired energy portfolio, a question that generates heated argument (Symons, 2019). Another question concerns the relationship between low carbon energy technologies and the competition for land. There are widely differing land intensities associated with different energy technologies (Figure 3.3). Nuclear energy has by far the smallest land footprint per unit of electricity generated, a full order of magnitude less than the conventional renewables of wind, solar, and hydropower and two orders of magnitude less than various forms of bioenergy. Many of these energy technology debates amongst reformed modernisers come to a focus around questions of ownership and governance (see Chapter 9). These are questions about

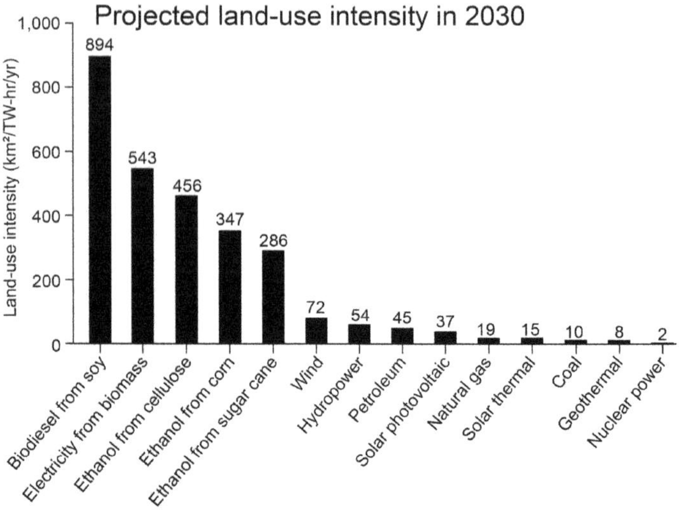

Figure 3.3 Land-intensity (i.e., how much land is needed to generate a unit of electricity) associated with various electricity generation technologies, estimated for 2030. The estimates consider both the footprint of the power plant, as well as land affected by energy extraction.
(*Source*: Redrawn from Hibbard et al., 2014).

the appropriate level of centralisation or decentralisation in energy systems and the different models and possibilities for democratic accountability that they entail.

Capturing and storing carbon

A parallel strategy to that of decarbonising the energy supply for some reformed modernists is the development of new techniques – or deployment of existing ones – for removing carbon dioxide from the atmosphere. The cluster of such techniques are usually referred to as carbon dioxide removal (CDR) technologies. These fall into two broad variants: Industry-based and nature-based solutions (Table 3.2). Industry-based examples include direct air capture, **carbon capture and storage**, and bioenergy with carbon capture and storage, so-called BECCS. These are sometimes referred to as "unnatural" climate solutions owing to their significant dependence on technological innovation. In contrast, nature-based solutions (NBS) involve the conserving, restoring, or enhancing of forests, soils, wetlands, grasslands, and

Table 3.2 Selected CDR technologies categorised according to a "natural–unnatural" binary classification.

"Natural" climate solutions	*"Unnatural" climate solutions*
Agroforestry	Bioenergy with carbon capture & storage
Biochar burial	Building with biomass
Coastal wetland restoration	Direct air capture and storage
Conservation agriculture	Enhanced weathering
Cropland nutrient management	Low-carbon concrete
Peatland conservation & restoration	Ocean alkalinity enhancement
Reforestation, conservation & management	Ocean fertilisation

(*Source*: Bellamy & Osaka, 2019).

agricultural lands as carbon sinks. These modified practices – some of them centuries-old – would either reduce carbon dioxide or methane emissions and/or remove carbon dioxide already in the atmosphere. One example is the controversial claim that planting over 1 trillion trees on 900 million hectares worldwide would sequester up to two thirds of the fossil carbon dioxide emitted in the industrial era (Bastin et al., 2019).

Some of the tensions that pervade debates about CDR technologies are similar to those we have seen previously in relation to energy decarbonisation. They are questions about the ownership and governance of technology and about the competition for land between different uses. For example, the Bastin et al. study of mass tree-planting has been heavily criticised not just for its carbon calculations but for entirely ignoring questions of land tenure, human livelihoods, and social justice. There are questions to be explored here that are very familiar for geographers. For example, what is meant by "natural" (Castree, 2013) and on what basis are NBS often valorised over techno-based CDR technologies? And who gets to decide what and whose land is being used for nature-based solutions?

Different versions of the reformed modernist vision are at stake here. For some, NBS for carbon dioxide removal appear cheaper, are already largely proven, and do not appear to present the same hazards to water use, biodiversity, and ecosystem services as do some industry-based CDR technologies. Indeed, it is often claimed that NBS "would bring additional benefits to ecosystem services, including to biodiversity, water filtration and flood control, soil enrichment and air filtration" (Bellamy & Osaka, 2019: 98). On the other hand, NBS provoke competition for scarce land and raise significant questions about justice and governance. NBS also elicit scepticism about the ease with which they can be scaled-up to deliver the magnitude of emissions reductions desired by reforming modernisers, a scepticism that also applies to industry-based solutions.

Reducing planetary heating

The modernist commitment to arresting climate change through technological innovation and development offers a further strategic option. The technologies discussed earlier either displace carbon in the energy mix or else capture carbon dioxide from the atmosphere. But technologies are also proposed by some modernisers that have the sole purpose of offsetting the additional atmospheric heating caused by greenhouse gases accumulating in the atmosphere. These technologies are often referred to as solar geoengineering. They offer the prospect of intentionally altering the Earth's radiative balance to restore climate to some prior condition or else to limit future climate change to some desired threshold. The most frequently considered proposal is to inject large quantities of particulate matter – often sulphate aerosols – into the stratosphere to mimic the cooling effect of volcanic eruptions. Although no such technologies have yet been deployed, theoretical calculations and computer simulations suggest that such technologies are feasible, relatively cheap to deploy, and effective in reducing incoming solar radiation.

Solar geoengineering is a climate management strategy broadly consistent with the presuppositions of the reformed modernist position. It relies upon science and engineering to be able to optimally design and deliver such interventions in the stratosphere. And it rests on the conviction that the technologies could be governed using the institutions of modernity (Reynolds, 2019; see Chapter 9). They are also technologies that can find their place within a reformed capitalism, offering significant new commercial and business opportunities. Some argue that solar geoengineering may well be necessary to deploy if the planetary temperature is to stay within the warming limits of 1.5° to 2°C embedded in the PACC (Long, 2020). Yet it is far from the case that solar geoengineering is welcomed by all climate reformers. The controversy resolves largely around the social, ethical, legal, and political regulation of the technologies (see Chapter 10).

Ecomodernism

Some of the tensions alluded to earlier within the broad stance towards climate change I describe as reformed modernism come into focus when one particular manifestation of this position is considered, namely ecomodernism.

The ideology of ecomodernism is most closely associated with the Breakthrough Institute, a non-profit organisation based in Oakland, California, established by Ted Nordhaus and Michael Shellenberger in 2003 and committed to the goal of "modernising environmentalism for the twenty-first century". This mission was first outlined in their provocative essay from 2004 titled "The Death of Environmentalism: Global Warming Politics in a Post-environmental World", published three years later in book form (Nordhaus & Shellenberger, 2007). Given their appropriation of "modernism" as a part-descriptor of their brand of environmentalism, it seems appropriate to place ecomodernists within the reformed modernism position I have outlined in this chapter.

Ecomodernism sets itself *against* those who claim that the emergence of climate change in the modern era reveals the fatal flaws of modernity (for more on this see Chapter 5). One of the founders of Breakthrough, Ted Nordhaus, argues just the opposite. "Can we [not] recognise", says Nordhaus, "that our contemporary projects of liberation and self-actualisation are *made possible only by* the extraordinary freedoms and abundance that industrial modernity made possible" (2019; emphasis added). Modernity itself is not the problem. Rather it is that modernity has not been fully realised. To tackle climate change effectively requires a reformed – or enlightened – modernity. As the Ecomodernist Manifesto of 2015 explains, humans need to use all their "growing social, economic, and technological powers to make life better for people, stabilise the climate, and protect the natural world" (Asafu-Adjaye et al., 2015: 7).

Ecomodernism shares many instincts in common with the ecological civilisation espoused in recent years by the Chinese Communist Party (see Box 3.2). It defines itself as adopting a morally progressive, optimistic, and pragmatic position. Through technological innovation and universal human development, ecomodernists deem it possible to arrest the most serious dangers of climate change and simultaneously benefit both humanity and the planet. They have a high view of science and technology and believe in the emancipatory possibilities of modernity, not least the right for all people to have access to reliable and cheap energy services that are essential for human well-being and dignity. They are called upon to embrace technological advances, to "Love your Monsters" (Latour, 2011: 1). In *Ecomodernism: Technology, Politics and the Climate Crisis* (2019), Jonathan Symons argues that ecomodernists are not simply techno-optimists. They are flag-bearers for social democracy, political pragmatists underpinned by a philosophy of universal human dignity. Ecomodernists see themselves as working incrementally to reform existing social and economic frameworks rather than to dismantle or transform them.

BOX 3.2 CHINA'S ECOLOGICAL CIVILISATION

A very different manifestation of the position on climate change I call here reformed modernism can be found in China. The idea of *shentai wenming* – "ecological civilisation" – has gradually gained ground in the Chinese Communist Party's ideology over the last 15 years. For example, in 2008 China's State Environmental Protection Agency was upgraded to the status of a Ministry and in 2017 was renamed the Ministry of Ecology and Environment. Between 2007 and 2017, more than 170,000 articles in mainstream Chinese media invoked the concept of ecological civilisation and in 2018 the concept was enshrined in the Communist Party's constitution. Nordhaus and Shellenberger sought to reinvent environmentalism for an American audience in the 2000s through their construction of ecomodernism. In a similar way, developing the idea of ecological civilisation could be seen as offering a distinctly Chinese response to climate change. Ecological civilization is seen as the final goal of social, cultural, and environmental reform within Chinese society. It argues that the changes to be wrought by a warming climate can only be headed-off through an entirely new form of civilisation, one based centrally on ecological principles. Its particular form seeks to be resonant with China's philosophical and institutional past and compatible with the centralised state bureaucracy of the present.

It would be wrong to interpret China's ecological civilisation narrative solely through the lens of Western notions of reformed modernity or ecomodernity. In its stance towards human rights and liberalism, China certainly differs from Western notions of modernity. Yet China's vision of ecological civilisation and the position of reformed modernism do hold some things in common. It is a vision partly inspired by the legacy of 1980s-style sustainable development, namely that environmental (and climate) concerns are as important as social and economic ones. It leans heavily upon a scientific diagnosis of the changing climate and is supremely techno-optimistic in its vision of how the condition of ecological civilisation can be achieved. In this sense it is very different from the more romantic version of eco-civilisation espoused by deep ecologists and new cultural movements such as the Dark Mountain Project, which seek "an unweaving of the core tenets of western civilisation" (see Chapter 5).

On the other hand, ecological civilisation complements the three core dimensions of sustainability with features that are specific to

China's political organisation and governance structure. Consonant with China's political culture, it envisions goal-directed progress through state command-and-control regulation within an authoritarian environmental governance paradigm. This is a very different model for governing climate to that offered by Western liberal environmentalism (Qin & Zhang, 2020). The Chinese state appears strong and omnipresent. Ideologically charged discourses, such as ecological civilisation, are intended to further reinforce its legitimacy. In reality however, the state needs a broad coalition of stakeholders to promote its vision. For China to realise an ecological civilisation it will rely heavily on cooperation from scientists, experts, the business community, private investors, and technological innovations for developing creative techno-designs for liveable and sustainable environments.

Read:

Hansen, M., Li, H. and Svaverud, R. (2018) Ecological civilization: Interpreting the Chinese past, projecting the global future. *Global Environmental Change*. 53:195–203; Kuhn, B. (2019) *Ecological Civilisation in China*. Berlin: Dialogue of Civilizations Research Institute.

Instead of harmonisation *with* nature, ecomodernists contend that human activities need to be locationally decoupled from nature to avoid economic and ecological collapse, so-called land-sparing. "Humanity must shrink its impact on the environment to make more room for nature" (Asafu-Adjaye et al., 2015: 6). Tackling climate change therefore presents an opportunity to spare nature and re-wild the Earth, to be advanced through urbanisation, more intensive energy generation and agriculture, and by embracing novel ecosystems. Since "pristine nature" and wilderness no longer exist, ecomodernists are receptive to rewilding, ecosystem novelty and new urban ecologies (Marris, 2011).

Ecomodernists share some convictions with the broad reformed modernism position outlined in this chapter, but there are some key differences. The two main lines of fracture are caused by its techno-optimism and its anthropocentrism. Ecomodernists are more optimistic than many about the potential for human innovation and ingenuity to manage technological progress for maximising well-being, whilst also sparing land and minimising waste. Technology sets humans apart from other species and for ecomodernists this anthropocentricism gives humans a special status with responsibilities towards the nonhuman world. This is not necessarily shared with all climate reformers.

With respect to the position of sceptical contrarianism outlined in the following chapter, ecomodernism is most clearly differentiated in that climate change is understood as presenting an important and pressing challenge. Even if there are perhaps some points in common between ecomodernists and lukewarmers (see Chapter 4), passivity or scepticism in the face of a changing climate is not an ecomodernist stance. And through its attitudes to technology, nature, humanism, and the state, ecomodernism is clearly differentiated with respect to the ecosocialism, green activism, and localism characteristic of transformative radicalism (see Chapter 5). Ecomodernists view negatively the desire for suburbanisation, rural existence, or small-scale and low-yield farming. For example, they would argue that low-yielding renewable energy technologies cannot supply the burgeoning and rightful energy demands of aspirational modernisation in the majority world, whilst at the same time sparing land for nature.

Chapter summary

This chapter has described some of the main components of a position on climate change I call "reformed modernism". For reformed modernisers, climate change is recognised as a consequence *of* modernity and yet also as a significant challenge *for* modernity. Yes, climate change is an outcome of 200 years of rapid fossil-fuelled technological expansion and accompanying growth in global population and economic activity. And, yes, the fruits and burdens of this modernity have been unevenly distributed across the surface of the planet. Yet the necessary responses to climate change are premised on a *reformed* modernity. Now is not the time for retreat or surrender. There should be no withdrawal from the impulses of technological advancement and human emancipation inspired by the modernist project. Neither is it a time to dismantle the institutions of modernity. Rather, it is by adjusting and redirecting the proven instruments of modernity towards more just and ecologically sensitive ends that climate change can be arrested. Reformed modernists advocate for incremental change rather than rupture, for reformation rather than revolution. Their argument is rooted in an enduring faith in the ideal of progress, bolstered by the realised achievements of modernity and its future promise. It is an ideology well-developed and illustrated in Stephen Pinker's book *Enlightenment Now: The Case for Reason, Science, Humanism and Progress* (2019).

A reformed modernism is one narrative storyline by which to establish the meaning of climate change. It is one that takes the science of climate change as offered by the IPCC at face value and then interprets what this scientific knowledge implies for political action. A reformed modernism should be seen as offering a "middle-ground" for how climate change can be tackled politically, a "third way" if you like (cf. Giddens, 1998). If this is so, then the following two chapters – outlining, respectively, sceptical contrarianism and transformative radicalism – offer the contrasting poles between which reformed modernists make their case. Both of these poles foreground climate science when establishing and justifying their political arguments about what climate change means for action in the world. So do reformed modernists. But as we shall see, the preferred visions of the future these three respective positions develop are conditioned on different claims about the status of scientific knowledge about climate change. Furthermore, while climate modernisers seek to *reform* the inherited institutions of modernity, climate contrarians and climate radicals adopt very different attitudes towards these institutions, especially those sustaining the organisation of world's political economy.

Notes

1 Carrington, D. (2020) Christiana Figueres on the climate emergency: 'This is the decade and we are the generation'. *The Guardian* newspaper on-line, 15 February, www.theguardian.com/environment/2020/feb/15/christiana-figueres-climate-emergency-this-is-the-decade-the-future-we-choose [accessed 1 July 2020].

2 Figueres, C. (2018) Limiting warming to 1.5C is possible – if there is political will. *The Guardian* newspaper online, 18 October, www.theguardian.com/environment/2018/oct/08/limiting-warming-to-15c-is-possible-if-there-is-political-will-climate-change [accessed 1 July 2020].

3 Branson, R. (2018) Why the energy revolution is one of the biggest opportunities of our time. *Virgin*, 1 May, www.virgin.com/richard-branson/why-energy-revolution-one-biggest-opportunities-our-time [accessed 12 August 2020].

4 Friedman, T.L. (2007) The power of green. *New York Times*, 15 April, www.nytimes.com/2007/04/15/opinion/15iht-web-0415edgreen-full.5291830.html [accessed 2 September 2020].

Further Reading

The case for a reformed modernism applied to climate change is nicely laid out in **Nick Stern's** *Why Are We Waiting? The Logic, Urgency and Promise of Tackling Climate Change* (MIT Press, 2015). Along similar lines also see **Anthony Giddens'** earlier book *The Politics of Climate Change* (2nd edn.) (Polity Press, 2011). **Paul Ekins** and colleagues have written a useful primer on the politics of energy in *Global Energy: Issues, Potentials and Policy Implications* (Oxford University Press, 2015), while **Axel Michaelowa's** *Carbon Markets or Climate Finance? Low Carbon and Adaptation Investment Choices for the Developing World* (Routledge, 2012) offers a good introduction to some of the economic instruments favoured by reformed modernists. **Bill Nordhaus** won the 2018 Nobel Prize for Economics for his pioneering work on adapting neoclassical economic theory to climate change and his book *The Climate Casino: Risk, Uncertainty and Economics for a Warming World* (Yale University Press, 2013) is a useful primer to his work. For a view challenging the efficacy of markets for achieving climate goals, then see **Danny Cullenward** and **David Victor's** *Making Climate Policy Work* (Polity Press, 2020). *India in a Warming World: Integrating Climate Change and Development* (Oxford University Press, 2019), edited by **Navroz Dubash**, is recommended as an excellent introduction to how a reformed modernist stance on climate change applies to a large developing nation such as India. In *The Power of Cities in Global Climate Politics: Saviours, Supplicants or Agents of Change?* (Palgrave Macmillan, 2018), **Craig Johnson** explores the potential role of cities in bringing about the incremental changes favoured by climate reformers. And two books to be recommended with a distinctly ecomodernist slant on the climate reformist position are **Dan Sarewitz's** *Climate Pragmatism* (CPSO, Arizona State University, 2017) and **Jonathan Symons'** *Ecomodernism: Technology, Politics and the Climate Crisis* (Polity Press, 2019).

QUESTIONS FOR CLASS DISCUSSION OR ASSESSMENT

Q3.1: The "green growth" paradigm rests on the conviction that the modernist pursuit of economic growth is compatible with arresting climate change. Do you agree with this conviction?

Q3.2: Examine and explain the range of numerical values that exist for the social cost of carbon (see Pezzey, 2019).

Q3.3: To what extent would appropriate worldwide pricing of carbon dioxide emissions provide a solution for climate change?

Q3.4: Could a fully renewable energy portfolio meet India's energy and development needs?

Q3.5: Why might reformed modernists be sceptical of nature-based solutions (NBS) to climate change?

Q3.6: If climate change is a manifestation of modernity, why do ecomodernists want to solve climate change through *more* modernity?

4

SCEPTICAL CONTRARIANISM

. . . Climate change contested . . .

Introduction

One of the most common metaphors used in public debates about climate change involves war. It is not that there is "a war" between humanity and climate change, as was implied for example when this metaphor was used in 2020 in the case of humanity versus the SARS-CoV-2 virus. Rather the war metaphor with respect to climate change alludes to the conflict between different *claims* about the reality and seriousness of the problem. Examples of the metaphorical language of this war include phrases such as "the climate wars", the "battle" between climate deniers and believers, the "fight" between green and neoconservative ideologies. The American climate scientist Michael Mann was able to title his autobiographical book *The Hockey Stick and the Climate Wars: Dispatches from the Front Lines* (2012) and everyone knew to what he was alluding.

This language of embattlement has been a peculiarly Anglophone framing of climate change in public discourse. Indeed, perhaps a peculiarly *American* framing – even if one that has been exported to some other countries. These war metaphors offer a simplified binary and confrontational version of what are in fact a multitude of public meanings of climate change. And yet it is a framing that has proven useful for both "sides" in the battle (Howarth & Sharman, 2015). It creates an "us" and a "them", a "with us" or "against us" mentality. It helps rally supporters and resources – both rhetorical and material – to a clearly defined cause. But it belies a much more complex and chequered landscape of private beliefs, public narratives, and political advocacy about the reality and significance of climate change, which this book is laying out.

In *Climate Change Scepticism: A Transnational Cultural Analysis* (2019), eco-critic Greg Garrard and colleagues revealed contrasting expressions of climate scepticism as found in different national and linguistic cultures. They showed that not everywhere is like America. Geography matters. Garrard and co-authors accomplished their task using the tools of literary criticism, more specifically those of **eco-criticism**. Using a variety of textual forms – blogs, novels, adverts, non-fiction – these authors dissected how climate scepticism is expressed culturally in four national settings, in France, Germany, the UK, and the USA.

Scepticism with regard to climate change was shown to exhibit very different motives and qualities in these four countries. Climate scepticism in France and Germany has much more in common than it does with UK and American scepticism. In Germany, scepticism presented as a critique of cultural pessimism, in France as a defence of secular humanism. British scepticism with regard to climate change manifested as more idiosyncratic than any of the other three cultures, reflecting a distinctly British style of libertarianism, cynicism, and comic nihilism. The USA displayed the most evidence of ideologically orchestrated climate scepticism. One important finding from Garrard's comparative analysis, considered in more detail later in this chapter, is the "near-universal complaint [of sceptical discourses] . . . that climate science contravenes the ideal of apolitical science" (Garrard et al., 2019: 238).

***

This chapter gathers together various interpretations of climate change that I label as "sceptical contrarianism"[1]. This position – as with reformed modernism in Chapter 3 – appears to be committed to a broadly modernist worldview. Science is valued as a powerful and authoritative cultural institution for producing reliable knowledge about the world. Rather than being anti-science – as is sometimes claimed – many forms of contrarianism in fact unite around a shared commitment to the cultural authority of science. What is recognised as "sound science" is called upon to form the basis for "sound public policy". It is for this reason that sceptical contrarianism earns its place in the book as a "science-based" position on climate change. Furthermore, sceptical contrarians would generally accept that the achievements of modernity – the nation state, individual liberty, technological progress, a mostly capitalist economy – are those that the world should continue to rely and build upon.

However, further investigation of contrarianism shows where the fault-line with the reformed modernists of Chapter 3 appears to lie. It seems to emerge around contrarians' claim that climate science has been corrupted by green ideology. Or at the very least that the dangers of future climate change claimed by bodies like the IPCC have been exaggerated. Rather than being "the most pressing issue of our time" requiring significant reform to modern institutions, climate change in the eyes of many contrarians becomes a relatively minor concern – or even for some a non-issue. Furthermore, what at first sight appears to be a distinction between reformers and contrarians – contrasting assessments of the veracity of climate science – in fact turns out to be rooted in different attitudes to risk, economics, the state, and modernity itself.

The chapter proceeds as follows. I first examine the nature of some of the challenges to climate science that emanate from sceptical voices. It is not that contrarians devalue science. Their critique is that the science of climate change has been unduly politicised. I then dissect the variety of contrarian convictions and dispositions with regard to climate change, revealing a complexity that exceeds the combative binary of "us" versus "them". There are many varieties and shades of climate scepticism and these each have different geographical, cultural, and political agencies. Science, values, culture, identities, and politics become profoundly entangled when constructing public meanings of climate change. Motives for climate contrarians include identity protection, ideological or religious conviction, emotional distancing, **cognitive dissonance**, economic rationalism, and libertarianism. The reasons for someone to hold a sceptical disposition with regard to climate change therefore extend far beyond the influence of a mendacious fossil fuel industry – as proposed for example by Oreskes and Conway (2010). Contrarianism cannot be defined by the hyper-partisan politics of the United States. The remaining subsections of the chapter then illustrate this complexity by fleshing out three distinct expressions of contrarianism: Passive scepticism, lukewarmism, and political populism.

There is considerable diversity within the broad interpretative position described in this chapter. The analysis of climate scepticism undertaken by Garrard and colleagues was undertaken with a close eye on cross-cultural differences. It is in this same spirit that the chapter – informed by the insights of *science and technology studies*, *eco-criticism*, *social psychology*, and *geographies of emotion* – outlines a geographically nuanced understanding of climate contrarianism.

Contesting climate science

The contrarian position on climate change is more complicated than being marked simply by a scepticism towards the findings and pronouncements of climate science. Many scientists and public commentators and advocates have mistaken one for the other, thinking that climate contrarianism is synonymous with scientific scepticism. Climate scepticism is believed to reveal a cognitive deficit within the individual with regard to scientific knowledge and their understanding of the phenomenon of climate change. As we shall see, this is not so. But for such scientists and advocates, it follows that contrarian beliefs must be expunged and behaviours changed through improved scientific literacy and/or through more robust communication of climate science. Relentless communicative emphasis on "the climate consensus" or the pursuit of "inoculation strategies" for protecting the public against scientific misinformation are regarded by some social psychologists as the essential tools for "defeating" contrarianism (Cook, 2020). (Note how we are back to war metaphors.)

This chapter challenges such a narrow understanding of contrarianism. But because climate contrarianism on the one hand and scepticism about climate science on the other are so often conflated, we need first to recognise some of the public criticisms that *are* brought against climate science and scientists by self-declared contrarians. These complaints do indeed feed the common misunderstanding that the essence of climate contrarianism is a contest about the credibility of climate science. Yet these criticisms are often not *anti*-science. In some cases, they may reveal a very strong commitment *to* science – or at least to a certain understanding of science. Contrarian critiques of climate science are often seeking to defend a set of ideals about science, ideals that need to be adhered to if science is to produce objective and reliable knowledge about the world. Three lines of argument illustrate the point.

A first set of accusations brought against climate science is that it reveals a corruption of "the scientific method". For example, climate science is seen as over-reliant on computer simulation models rather than rooted in observed data. Modelling is perceived as a sub-scientific method for acquiring truth, one that fails to conform to the three standard scientific precepts of theory, experimentation, and data. Or take the complaint about the climate consensus. Critics will say that climate science, notably the IPCC, has been led – whether willingly or not – into forming a premature consensus on what is known about the effect of human actions on the world's climate. Again, the contrarian argument runs that objective scientific truth is never arrived at through such fallible and subjective processes as consensus-making. Or take a third complaint which is directed against the individual scientist. The ethos of the climate scientist has been compromised by the politicisation of climate science, alluding to the archetype of the "Corrupted Scientist" (Cloud, 2020). One evident source of this complaint was the **Climategate** controversy, which erupted late in 2009 when leaked emails between leading climate scientists were claimed to reveal dubious – if not illegal – scientific practices. This complaint may not have been wholly without warrant, as science studies scholar Ramírez-i-Ollé observes: "Scientists and their critics alike interpreted the stolen emails as embarrassing deviations from the alleged social demands of a consensual, objective, and accessible science" (2015: 384).

To explain these perceived corruptions of science, critics then turn to a second set of arguments, which are about the funding of climate science and about ideologically motivated alarmism. Far from erring on the side of *least* drama, it is claimed the IPCC has come under sustained pressure to *exaggerate* its findings. Climate science has become overtly politicised and is infected by an ideology of environmental alarmism about the future. The contrarian science columnist Matt Ridley expresses this accusation in blunt terms. "Green multinationals, with budgets in the hundreds of millions of dollars, have now systematically infiltrated science", Ridley says. The result is that "many high-profile climate scientists . . . have become one-sided cheerleaders for alarm, while a hit squad of increasingly vicious bloggers polices the debate to ensure that anybody who steps out of line is punished" (2015: 4).

These two lines of complaint – that climate science fails in its ethos of disinterestedness and that this is because of ideological contamination – lead contrarians to pursue a third mode of intervention. If climate science has been corrupted through subjectivism, ideological bias, and politicisation, then correctives are necessary to reclaim the objectivity of true science. Thus, it is necessary for public auditors to hold climate science and scientists to account. It is necessary to reassert the normative value for the scientist of a sceptical disposition, namely scepticism as "doubt or incredulity as to the truth of some assertion or supposed fact" (Oxford English Dictionary). For this reason, Andrew Montford titled one of his critiques of climate science *Nullius in Verbia: On the Word of No-One* (2012), appropriating this ancient motto from the UK's Royal Society in order to reclaim and revalorise scepticism as scientific virtue.

Similarly – and as we saw in Chapter 2 – the Heartland Institute mimicked the IPCC by publishing a series of climate reports under the auspices of the Nongovernmental International Panel on Climate Change, the NIPCC (Figure 4.1). And in Matt Ridley's pamphlet "The Climate Wars and the Damage to Science" (2015) – written for the contrarian UK think-tank the Global Warming Policy Foundation – he expresses his concern about the lasting damage to the reputation of "science in general" that this corruption of climate science has engendered. Ridley laments that "it risks destroying, perhaps for centuries to come, the unique and hard-won reputation for honesty which is the basis for society's respect for scientific endeavour. . . . Science will need a reformation" (p. 14). This self-appointed campaign to detoxify climate science is what lies behind claims from high profile contrarian politicians to regain "sound science" as the basis for policy making. This explains, for example, Alaskan governor and Tea Party co-founder Sarah Palin's remark from 2009 (italics

NIPCC

Nongovernmental International Panel on Climate Change

Figure 4.1 The logo of the Nongovernmental International Panel on Climate Change, a series of reports published by the Heartland Institute in Chicago, mimicking the IPCC.

mine): "I've always believed that [climate] policy should be based on *sound science*, not politics".[2]

How are we to understand such criticisms of climate science and of the IPCC? At face value they could be taken as examples of holding science to account, of offering "constructive criticism". This would uphold what Ramírez-i-Ollé (2018) refers to as "civil scepticism" being a valued norm of scientific discourse. And it would undoubtedly be how the contrarian Canadian climate blogger Steve McIntyre would justify his blog "Climate Audit", which he started in 2005. McIntyre would claim he has been conducting a public *audit* of climate science. And many science and technology studies scholars would agree with the potential benefits of sceptics' critical perspectives. Thus, Myanna Lahsen concluded her study of American climate scientists by arguing that "societies may be well served by some level of diversity of opinion beyond that which exists within the IPCC, in line with democratic theory . . . and with the anthropological principle of societal resilience through diversity of perspectives" (Lahsen, 2013: 749).

At another level, however, these complaints reveal a limited and idealised understanding of how science works in practice, what Marianne Ryghaug and Tomas Moe Skjølsvold (2010: 304) refer to as a "kind of Boy Scout image of science". And at a third level, one might argue that they reveal a malicious and disingenuous attack on science, the deliberate sowing of public doubt about the credibility of climate science for reasons of political or economic self-interest. Naomi Oreskes and Eric Conway's account in *Merchants of Doubt: How a Handful of Scientists Obscured the Truth on Issues from Tobacco Smoke to Global Warming* (2010) of ideologically motivated denialism of science is pertinent here and not without merit. All three of these interpretations may be valid to various degrees. There are different motives at work amongst different complainants. By no means are all climate sceptics in the pocket or pay of fossil fuel interests. And we need to remember that *all* scientists have their motivations beyond simply a desire to "discover facts", whether it be to become famous, gain tenure, win awards, or be well-paid. Disinterestedness is an ideal of science as a social institution rather than it being a virtue that any individual scientist can consistently adhere to.

However, the arguments illustrated previously that frequently erupt over the integrity of climate science are often propelled by wider political struggles to control the public meaning of climate change. In other words, to control the public narrative. In this task, science is regarded as a crucial cultural and political resource that is important to have on one's side. If one can bring scientific evidence into alignment with one's ideological outlook and political preferences then one's public policy position appears strengthened. The reality is, however, that the social meanings of climate change are rooted in beliefs, values, ideologies, and identities. They are not rooted in scientific evidence. These motivational commitments are used to *interpret* scientific evidence, to pass judgement on the facts so that they appear in alignment with one's values. This is as true of reformed modernists and transformative radicals as it is of sceptical contrarians.

Varieties of contrarianism

The attention placed on the mendacity of the fossil fuel lobby in the United States and elsewhere – the villains of Exxon Mobil and the Koch brothers – has been influential in the public framing of climate denialism (Oreskes & Conway, 2010; Dunlap & Jacques, 2013). But at best this offers only a partial and rather superficial explanation for many of the sceptical contrarian understandings of climate change I am exploring here. It also falls into the danger of American exceptionalism, the danger of reading into other national and cultural contexts the motives and political dynamics that prevail in American public culture. It is important to understand *why* climate science may be challenged or contested by people in a variety of geographical settings. A different vocabulary is needed in order to move away from the erroneous conflation of climate contrarianism with a sceptical attitude towards climate science. A richer and more differentiated language is required than is offered by the binary option of either "mainstream" or "contrarian", of "us" versus "them" (Howarth & Sharman, 2015).

There are a number of ways in which this vocabulary can be enriched. One common and useful move is to differentiate between so-called trend, attribution, and impact scepticism. By this is meant, respectively, is climate change happening, is it human-caused, and will it have a serious impact? Taken together these three positions define *evidence scepticism*, expressing doubt about what is known. This is to be differentiated from *response* or *policy scepticism*. This descriptor captures those who question not whether climate change is happening or whether it is human-caused – this is accepted as true – but whether the policy responses to the challenge of climate change are appropriate or feasible. Response sceptics question what should be done, how, and by whom (Poortinga et al., 2011). But these categories of evidence and response scepticism don't get us much further forward than the rather superficial caricature of a "classical climate sceptic", nicely expressed by Bob Henson in his book *A Rough Guide to Climate Change* (2006: 272):

> The atmosphere isn't warming; and if it is, then it's due to natural variations; and even if it's not due to natural variation, then the amount of warming will be insignificant; and if it becomes significant, then the benefits will outweigh the problems; and even if they don't, technology will come to the rescue; and even if it doesn't, we

shouldn't wreck the economy to fix the problem when many parts of the science are uncertain.

This is still too simple. It doesn't begin to explain *why* people hold these views, or recognise how and why these views may *vary* from place to place. For that, more careful qualitative studies are needed to reveal the full complexity of how people think, feel, and act in response to climate change. One example of such study comes from Australia, in which a representative sample of about 100 citizens from Canberra and surrounding districts were selected for a series of deliberative discussions (Hobson & Niemeyer, 2013). Around 40 per cent of this group disclosed some level of doubt or scepticism about climate change. But rather than being a homogenous group captured by the label "climate sceptic", these questioning citizens revealed a variety of viewpoints and attitudes about climate change (Table 4.1). Grouping them all together under one contrarian label would obscure the reasons why people held the views they did.

Useful insights into these questions can also be gained from the results of the Yale Program on Climate Change Communication. Over the last decade this program has been surveying American citizens' beliefs and attitudes to climate change. They

Table 4.1 Five discursive positions on climate change held by Australian citizens.

Descriptor	*Explanation*
"Emphatic Negation"	given the current state of knowledge about climate change, nobody is in a position to say whether it is real or not
"Unperturbed Pragmatism"	a phlegmatic optimism that society will be able to adapt to any coming changes; there is still time to adapt
"Proactive Uncertainty"	a relative indifference to policies that address climate change in a fairly broad sense and thus a tendency towards inaction
"Earnest Acclimatisation"	climate change as a natural phenomenon, about which we should be concerned; since humans have not contributed to causing it, we should adapt rather than reduce carbon emissions
"Noncommittal Consent"	climate change is probably anthropogenic, but given the uncertainty the emphasis should be on tackling potential impacts rather than causes, adaptation is more important than mitigation

(*Source*: Hobson & Niemeyer, 2013).

have been able to segment the national population according to the following six dispositions: The "alarmed", "concerned", "cautious", "disengaged", "doubtful", and "dismissive". This result has prompted the study to be labelled Global Warming's Six Americas (Figure 4.2). Similar population-level survey work in India by the same program has offered a slightly different segmentation given India's different political and cultural context: The "informed", "experienced", "undecided", "unconcerned", "indifferent", and "disengaged". Such deliberative and survey-based investigations of how people in different national settings think about climate change can be set alongside studies like Garrard's cross-cultural exploration of climate scepticism. Together, they allow the possibility of mapping a more complex cultural geography of sceptical contrarianism than is suggested by the casual use of the othering and universalising label "climate denier".

Passive scepticism

Hobson and Niemeyer's study from Australia shows that there are many different ways in which people question the reality and significance of climate change. The beliefs of "Emphatic Negators" are very different from those of "Earnest Acclimatisors". And as Yale's Six Americas studies show, most Americans' attitudes to climate change – around 60 per cent in the most recent

Figure 4.2 Representations of the six population segments identified in Yale's Six Americas study.
(*Source*: Michael Sloan).

2019 survey – fall somewhere *between* the extreme positions of being either "alarmed" by climate change or "dismissive" of it. Similarly, for India. The majority of people in these two countries revealed to be "cautious", "disengaged", "indifferent", and "doubtful" express forms of sceptical contrarianism that need to be recognised and understood. Grasping the varieties and geographies of contrarianism is important.

Valuable for this purpose are the insights of sociologists and anthropologists gained from studies of individual, social, and cultural dispositions. What is important are the *reasons* that people give for their sceptical contrarianism with regard to climate change. These may include a general distrust in environmental science, a wariness about environmentalism as an ideology, a reluctance to change behaviours, or an apathy or fatalism brought about by feelings of helplessness (Poortinga et al., 2011). Others may simply be disinterested or bored by the topic or find that the struggle for everyday survival squeezes out concerns about more distant threats. This latter may very often be the case for people whose livelihoods are insecure or who feel they have little stake in the future.

These self-disclosed reasons take us well beyond accounts from (mostly) American social scientists of ideological denialism infecting American public culture and of individuals' cognitive deficits. People are not always the unwitting victims of other people's propaganda campaigns or ideological biases. Rather they are active in forming and expressing their beliefs rooted in their own life histories. These beliefs originate in active cognitive, social, and experiential processes and become inscribed in people's identities and sense of place. Beliefs, knowledge, experience, and behaviours are mutually reinforcing. Geographers' attention to the role of place, culture, and identity in shaping people's understanding of climate that we examined in Chapter 1 is important here. And we will return to this again in Chapter 6 when considering subaltern voices and in Chapter 8 with regard to religious engagements. For now, a few examples of more place-centred ways of understanding sceptical contrarianism are offered.

One of these comes from Kari Marie Norgaard's anthropological study of a small Norwegian town (see Box 4.1). Norgaard goes beyond thinking of climate change scepticism as an individual cognitive matter and shows how it is socially organized and reproduced. Another example is the study of "unconcern" about climate change conducted by Chloe Lucas amongst residents of Hobart, Tasmania. For Lucas, it was important to pay careful attention to the experiential conditions and life stories through which concern about climate change is resisted (Lucas & Davison, 2019). Their study of a small group of Hobart's citizens who self-declared to be only a "little concerned" by climate change therefore used the technique of narrative inquiry. Narrative inquiry elicits people's representation of their life experience and identity in the form of stories. Lucas and Davison were able to analyse the stories of unconcern about climate change disclosed by their Hobart citizens. Evidence was forthcoming for the five main explanations for lack of climate change concern identified in the research literature: Ideological, group-based, religious, self-enhancing, and self-protective. They summarised it thus:

> Rachel's unconcern about climate change is self-protective, associated with group-based responses and, to a lesser extent, religious beliefs. Gerald's unconcern is also focussed on the rights and interests of his social group and thus brings in self-enhancing and ideological responses (although his politics are ambiguous), while Doug's unconcern also appears to be group-based, self-protective and, to a lesser extent, ideological.
>
> (p. 142)

The reasons for unconcern about climate change expressed by these citizens are rooted in focal life concerns, which appear largely incommensurable with the dominant narratives of climate change prevailing in public discourse. These Australian citizens express passive scepticism about climate change in terms far removed from the accounts of climate denialism which denounce nefarious fossil fuel interests. As Lucas and Davison explain, the participants in their study were not "unconcerned about climate change because of any inability to comprehend sophisticated knowledge, slavish obedience to tradition or authority, deliberate indifference to the suffering of others, or narcissistic self-interest" (p. 145). Neither, they claim, should people who are relatively unconcerned about climate change be assumed to be pursuing a less moral life than those who are campaigning on the streets. People's level of concern about climate change emerges from the lived contexts of their lives, communities, and larger ideational attachments. It cannot easily be extracted from the attitudes or attributes of individuals revealed in large-scale anonymous surveys.

BOX 4.1 THE NORWEGIAN TOWN OF BYGDBY – *LIVING IN DENIAL* BY KARI NORGAARD

In *Living in Denial: Climate Change, Emotions, and Everyday Life* (2011), American social anthropologist Kari Marie Norgaard explored the sociological construction of climate scepticism. Norgaard lived for 12 months in a small rural Norwegian town, to which she gave the pseudonymous name of Bygdby. There, she observed and participated in cultural activities – such as sheep slaughtering and collective story-telling – and listened to the hopes and fears expressed in this

(Continued)

unassuming small-town community. Norgaard showed how emotions are kept distant from everyday life, especially among Norwegian men. Cultural norms of masculinity involve being strong and the men of Bygdby rarely discussed climate change in emotional terms, except in gatherings where alcohol was involved. Through this ethnography, Norgaard paints a picture of how a modest rural Norwegian society engages with the idea of climate change. She explores the interactions between scientific evidence, personal experience, collective belief, and cultural practice. It produces what she called "the social organisation of climate change denial". By this she means that whilst scientific information about climate change is known in the abstract, these facts are disconnected from political, social, and private life. She helps us better understand the cultural constraints that lead to political quietism concerning climate change – the *absence* of protests, social activism, and public action.

Norgaard's study offers a rich and textured illustration of two important truths about how the idea of climate change works in the human world. The first is that the meaning of a scientific fact is not for science to define. Scientific evidence is always contextualised and interpreted through cultural filters, what Dan Kahan – building upon the work of anthropologist Mary Douglas – has called "the cultural cognition of risk" (Kahan et al., 2011). Citizens make sense of climate change for themselves, rather than simply imbibe what scientists say climate change is or what it means. The second truth is that people find it very hard to engage imaginatively and emotionally with largely invisible and globally mediated risks like climate change. In this respect, Norgaard's study is valuable for her deep emphasis on "the feelings that people have about climate change and the ways in which these feelings shape social outcomes" (p. 210).

For many citizens, the problem of climate change is not really climate change. Rather, climate change becomes a synecdoche. It stands in for something else, for a sense of unease or anxiety about the contemporary world and its future. As we saw in Chapter 1, the idea of climate change unsettles presumed cultural, social, and environmental certainties. And when struggling to find the meaning of climate change, people discover contradictions emerge between their knowledge, their values, and their actions. As the people of Bygdby realise, there is no easy way of resolving such cognitive dissonance.

Read:

Norgaard, K.M. (2011) *Living in Denial: Climate Change, Emotions and Everyday Life*. Cambridge, MA: MIT Press; See also Carolan, M. (2010) Sociological ambivalence and climate change. *Local Environment*. 15(4): 309–321.

According to Norgaard (Box 4.1), sceptical contrarianism is often active but rarely is it persistently conscious. Many individuals struggle to keep climate change distant from everyday life, despite living in conditions where the manifestations of climate change may erupt any time. This is similar to the notion of *sociological ambivalence* – or the related psychological idea of *cognitive dissonance*. People experience cultural roles and expectations, social obligations, and the tacit power of ideologies, in ways that often pull them in different directions. Most of us often have "mixed feelings" about people, events, or things in general. We hold incompatible or contradictory normative expectations of our attitudes, beliefs, and behaviours. Managing these mixed feelings is part of how people make sense of the world around us. The idea of climate change brings this ambivalence into the foreground. In his study based in the Colorado town of Fort Collins, Michael Carolan (2010) shows how ordinary Americans navigate these feelings of ambivalence with regard to the idea of climate change. This is captured well in the response of one citizen:

> Yes, I think climate change is real. And, yes, I think humans are ultimately to blame. And, don't get me wrong, I worry about it. I mean, I basically think my kids and their kids are going to be screwed if we don't do something quick. . . . Then there is the question of what can I do? I mean, I know what I can do – reduce my carbon footprint and stuff like that. But it's immensely discouraging knowing that you'll never see any positive net effects from any individualistic act. One person can clean a stretch of road and see the results. With the climate, well, that's a different story.
>
> (p. 315)

And another citizen offers their view on managing this ambivalence: "[climate change is] overwhelming. How couldn't it be? It's something that has the potential to impact everyone. . . . I could either lie awake at night thinking about it, which, honestly, isn't going to change anything, or I don't. I've chosen the latter" (p. 316). Rather than scepticism being an exceptional condition, there is a sense in which the disorienting effects of the idea of climate change makes all of us sceptical contrarians.

Lukewarmism

Norgaard and others identify the sociological construction of climate denial as a form of passive scepticism. But there are other expressions of sceptical contrarianism that are more active and overtly political. We will look briefly at two of these in the remaining subsections of this chapter: Lukewarmism and populism. Lukewarmism complicates simplistic associations of contrarianism being instinctively anti-science, whilst the phenomenon of political populism offers a very different lens through which to view climate contrarianism.

The label "lukewarmism" was coined in the Anglophone world in the later 2000s and became visible in public discourse during the following decade (Norton & Hulme, 2019). Those who claim this moniker would see themselves as valorising the achievements of science but at odds with the conventional interpretation of climate science. They therefore distinguish themselves from other contrarians who may be perceived to be denying the entire science of climate change. Lukewarmers agree that carbon dioxide is a greenhouse gas, that the world is warming and that a significant fraction of this warming is down to humans. But they exhibit mild forms of both evidence and response scepticism. As we saw earlier in the chapter, lukewarmers see themselves as adopting a healthy scepticism with regard to climate models and their predictions of the future. They contend that the IPCC process is flawed due to ideologically motivated scientists using evidence selectively. Lukewarmists recognise that climate change represents a risk to society worthy of a response, but they remain unconvinced that this risk is substantial or that its impacts would be catastrophic.

Lukewarmism is one way of triangulating between the polar categories of denialism and alarmism. It is well characterised by the Danish environmentalist Bjørn Lomborg (see Box 4.2). Even though global warming might be understood by lukewarmers as bringing some benefits to the world – for example that it has saved the world from another ice-age – they nevertheless advocate for modest climate policies. These policies typically promote adaptation to climate change over aggressive mitigation. Economic growth is necessary to ensure future prosperity. And they are optimistic that this prosperity, combined with human ingenuity, will enable successful adaptation to future climatic hazards. Lukewarmers argue that future investment in, for example, improved flood defences, plant breeding, and climate-proofed infrastructure will offset many of these risks. And although not opposed to renewable energy, such technologies are not considered sufficient or suitable to fully meet future energy requirements. Nuclear and clean coal technologies are favoured.

BOX 4.2 BJØRN LOMBORG AND THE COPENHAGEN CONSENSUS

The Danish environmental economist Bjørn Lomborg first came to international prominence in 2001 with the publication of *The Skeptical Environmentalist: Measuring the Real State of the World* (Cambridge University Press, 2001), an English translation of his 1998 Danish original. When published in English, *The Skeptical Environmentalist* received excoriating reviews from some established climate scientists and many environmentalists and prompted a threatened boycott of the prestigious publishing house Cambridge University Press. Several environmental scientists brought complaints of scientific dishonesty against Lomborg to the Danish Committee on Scientific Dishonesty, an official governmental body. The outcome of these complaints was confused, contested, and generally inconclusive for all parties concerned. A few years later, Lomborg published *Global Crises, Global Solutions* (2004), in which eight leading neoclassical economists were invited to present their arguments about how best to spend a hypothetical $50 billion to benefit humankind. Those policies most closely associated with conventional climate mitigation came low on the list.

Although the moniker "lukewarmer" did not exist in the mid-2000s, Lomborg's interpretative analysis of climate change matches the description. He takes a fairly mainstream view of climate science but argues that climate change is not necessarily the world's greatest problem. He stresses the need to decarbonise the energy sector but that such ambitions should be tempered by ensuring that there is sustained investment in public health, literacy, and infrastructure programmes to meet the human welfare needs of the world's poor. Lomborg's views on climate change – as with other lukewarmers – should be understood as manifesting the logic of a particular style of economic rationalism, development ethics, and cost–benefit **utilitarianism**. This logic separates him from reformed modernisers, even though he and other lukewarmers are sometimes associated with ecomodernism (see Chapter 3).

Bjørn Lomborg is a useful person to highlight because in debates about climate change he often serves as a symbolic "cutting-gate", the mechanism used by ranchers to divide cattle into separate pens. But if the full scope of the idea of climate change is to be grasped, then sceptical contrarians – whether Lomborg, lukewarmers in general, or the citizens of Bygdby – deserve to have their arguments and emotions listened to and engaged with. Such views should not be dismissed through personal attacks nor trivialised by naïve assumptions about cognitive or concern deficits. Nor should such individuals be blacklisted through derogatory labelling. Democracies require not less argumentation but more – and better-quality – arguments. As Taylor Dotson argues in *The Divide: How Fanatical Certitude Is Destroying Democracy* (MIT Press, 2021), democracy cannot function if citizens come to believe that protagonists of only one side of a debate are capable of reasoning rationally.

Read:

Lomborg, B. (ed.) (2010) *Smart Solutions to Climate Change: Comparing Costs and Benefits*. Cambridge: Cambridge University Press; Lomborg, B. (2020) *False Alarm: How Climate Change Panic Costs US Trillions, Hurts the Poor and Fails to Fix the Planet*. New York: Basic Books.

Richard Douglas (2018) dissects the lukewarmist position in an interesting analysis of the rhetorical claims of what he calls "environmental scepticism" but broadly similar to the position I describe earlier. Drawing upon climate-sceptical texts from individuals and think-tanks around the world, Douglas groups the central propositional claims of lukewarmists according to the rhetorician's tropes of ethos, pathos, and logos – in other words the authority, persuasiveness, and logic of speech (Table 4.2). Douglas' analysis is valuable in that it reveals the positive appeal of lukewarmism, which defends a modernist worldview and a particular view of science. In contrast to the modernists of Chapter 3 who understood climate change as necessitating a reform of modernity, lukewarmists seek to protect the core institutions of modernity from what they perceive as an attack upon them conducted in the name of climate change.

Political populism

There is one further manifestation of the sceptical contrarianism position to briefly look at. It is one that has become more visible and assertive over the last decade and is associated with the political populism that has recently found traction in a number of significant democracies around the world. The recent surge in political populism has been a reaction to the rising globalism of the last 30 years, itself rooted in the post–Cold War neoliberal doctrine of the **Washington Consensus**. Globalism envisions a world moving inextricably towards the adoption of a unified set of rules and standards – in economics, politics, and international relations. National borders gradually lose relevance and even disappear. Cultural distinctions give way to universal liberal values. Electoral democracy and free-trading capitalism spread the world over. Eventually, all countries become governed in more or less the same way.

Populism is a rejection of this cosmopolitan vision. In Western democracies, populism may be understood as a rejection of an elite establishment consensus that has benefitted enough people to count but too few to matter. Populist politicians speak for those who have felt excluded from conventional

Table 4.2 Positive appeals made by environmentally sceptical rhetors.

Ethos *Appeal to character* *The authority of speech*	*Pathos* *Appeal to emotion* *The persuasiveness of speech*	*Logos* *Appeal to rationality* *The logic of speech*
Reason and real scientific method tell us the truth about the world	Liberty is the highest good	You can't stop economic progress
Common sense and our experience of history are the best guides to the future	The point of politics and economics should be to increase the material welfare of the highest number of people	There are no limits to humankind's ingenuity and problem-solving abilities
Ordinary people understand life and morality better than liberal elites		The market coordinates individuals in a way that enables us to adapt to changing conditions and to solve complex problems
		The world is actually getting better and better

(*Source*: Douglas, 2018).

politics and who have not seen the benefits of globalism. Amidst the various ways of defining populism, two key markers would be these. First, that nothing should constrain the sovereign will of the "true people" of a polity and, second, that "we, the people" are locked in conflict with "outsiders", whether these outsiders be elites or immigrants. In the words of one political theorist, populism is the idea that "the employment of political power ought to be governed directly by the views of . . . the *populus* ['the people'], unmediated and unencumbered by social elites" (Sharon, 2018: 360). Populism usually manifests in offering simple answers to complex problems, often appealing to chauvinistic prejudice over credible evidence. Populism is predominantly associated with the political Right,[3] although in countries such as India or Indonesia populism would better be described as ethno-nationalist. Examples of populist leaders elected in recent years include Jair Bolsonaro (Brazil), Narendra Modi (India), Joko Widodo (Indonesia), and Donald Trump (the USA).

In contrast to a world projected through the ideology of globalism, maybe the world of the future is more like one described recently by an international team of climate change social scientists. Amongst their family of five future worlds – so-called socio-economic pathways – is one labelled "Regional Rivalry: A Rocky Road" (O'Neill et al., 2017). It is described thus:

> A resurgent nationalism, concerns about competitiveness and security, and regional conflicts push countries to increasingly focus on domestic or, at most, regional issues. Policies shift over time to become increasingly oriented towards national and regional security issues. Countries focus on achieving energy and food security goals within their own regions at the expense of broader-based development. Investments in education and technological development decline. Economic development is slow, consumption is material-intensive, and inequalities persist or worsen. Population growth is low in industrialized and high in developing countries. A low international priority for addressing environmental concerns leads to strong environmental degradation in some regions.

The argument to be made here is that many Right and far-Right anti-liberal populist parties and their followers display epistemic and/or policy scepticism with regard to climate change (Forchtner, 2019). Such populist contrarianism might express itself in different ways. One extreme reaction is to dismiss climate change as a hoax. Less extreme views might include the claim that climate policies are imposed on the ordinary citizen by cosmopolitan and liberal elites or that such policies are in any case counterproductive in environmental terms.

These forms of populist scepticism may erupt for a complicated mix of reasons. For example, President Macron's introduction of carbon taxes in France in 2018 led to widespread popular uprisings. As we saw in Chapter 3, carbon taxes are a favoured instrument of reformed modernists for tackling climate change by internalizing the environmental costs of carbon emissions. But the reaction of the French *gilets jeunes* – "the yellow vests" – to such taxes was violent street protests. This uprising of "the people" drew political energy from the argument that carbon taxes were a regressive climate policy. It was seen to be imposed by the political elite against the people, especially against those who had been "left behind" by the processes of globalisation. These protesters were drawn from both the political Right *and* the political Left. The sense of a president out of touch with the ordinary people is a common sentiment that feeds national populism. This indeed happened in 2018 and 2019 for France's right-wing political party – *Rassemblement National*, "the National Rally" – as they made significant gains against Macron in the May 2019 European election.

There are other reasons why right-wing populism may express itself in sceptical contrarianism. There is a quasi-latent environmental tradition within European conservatism – and in those further to the extreme Right, as in Nazism – due to the significance attributed to the link between "the land" and "the people", between nature and nation. This may be understood as offering an ethno-nationalist imaginary in which "the people" are rooted in a particular place. This sentiment is captured in the German term *heimat* – "homeland". For such conservatives, people are deeply embedded in nature, not separated from it.

One recent example of this appropriation again comes from France. In 2019 the leader of the National Rally in France – Marine Le Pen – reclaimed Nazi-era blood-and-soil rhetoric from the 1930s.[4] In the run-up to European elections, she offered a pledge to make Europe the "world's first ecological civilization" (see Box 3.2). Le Pen drew a distinction between the "ecologist" social groups who are "rooted in their home" – the people of the soil – and the "nomadic" people who "have no homeland" and "do not care about the environment" – the rootless elite. This creed of nature-nation-racial purity is not just a legacy of a certain strand of continental European thought. It is also present in the Anglophone world. This has been seen, for example, in discussions around overpopulation and environmental protection in relation to the "othering" of the immigrant. Political conservatives resist the rootless cosmopolitan environmentalism often expressed by the political Left through its concern for global climate (Scruton, 2012). Right-wing populists' commitment to the environment is focused on maintaining national sovereignty over their traditional communal homeland – their *heimat*.

The ethno-nationalist political Right might also appropriate for their own ends the rhetoric of "climate emergency" put into play by climate radicals (see Box 9.3). They may seek to advance far-Right policy programs by pursuing policies that can be justified and implemented under conditions of emergency. This danger of a "reactionary modernism" was pointed out more than 20 years ago by Meera Nanda in her study of Hindu nationalism (Nanda, 2003). A more recent version of this argument has been put forward by Nils Gilman of the Berggruen Institute in California. Gilman terms this form of climate contrarianism "avocado politics" – green on the outside, brown(shirt) on the inside (Gilman, 2020). This metaphor draws upon the earlier idea of "watermelon politics" – green on the outside, red on the inside – elaborated in James Delingpole's *Watermelons: How Environmentalists are Killing the Planet, Destroying the Economy and Stealing your Children's Future* (2012). For its detractors, watermelon politics repackaged the wish list of the political Left as environmental policies. Avocado politics would do the same for the political Right. In avocado politics there emerges a curious alliance between right-wing environmentalism and climate policy radicalism.

Chapter summary

The interpretative positions on climate change explored in this chapter are grouped together under the label "sceptical contrarianism". To explain contrarianism, I have largely favoured a sociological, cultural, and geographical analysis over a cognitive or political one. Contrarianism may be characterised as displaying an absence of "appropriate" concern about climate change and is variously described by others as climate apathy, quietism, ignorance, scepticism, or denial. These dispositions are often understood as being characterised by an opposition to climate science – or even to science in general. Or, at the very least, characterised by the appearance of an overly casual attitude to the scientific reasons for concern about the climatic future. The logic of this reasoning leads some academics and public advocates to develop campaign strategies to eradicate the perceived cognitive deficit of such contrarian members of the public. Or alternatively to expose the ideological biases of those political actors who sow public doubt or misinformation about climate change.

But this is to misread the range of political, cultural, social, and psychological reasons why people hold contrary beliefs – or else remain ambivalent – about the reality or the seriousness of climate change. A perceived "concern deficit" (Lucas & Davison, 2019) cannot be remedied simply through corrective political, education, or communication programmes. Arguments about the validity of the evidential claims of climate science stand-in for wider disputes about modernity, elitism, culture, and identity. In a broad sense then, contrarianism could be seen as defending a modernist or – in the case of political populism – a culturally conservative worldview against an attack on modernity's foundations launched by climate radicals (as we will see in Chapter 5). The disjuncture between contrarians and radicals is not fundamentally about science, nor even really about everyday politics. The disjuncture operates at a deeper philosophical level about *who* people think they are, *how* they think they should live their lives, their sense of empowerment, and their hope for the future. The answers to these questions are influenced strongly by geographical considerations, which are engaged with more explicitly in Chapters 6, 7, and 8.

Sceptical contrarianism is presented in this chapter as an example of a "science-based" position on climate change. But climate contrarianism, properly understood, is rarely about disputing the details of climate science. The arguments that erupt around climate science should be understood as part of a struggle to secure the cultural authority of science as legitimation for the preferred political response to climate change. Very much as the various European combatant nations in the First World War declared "God on our side", so too with climate change. None of the three positions outlined in Chapters 3, 4, and 5 can afford to dispense with science as an ally.

And yet as Raul Lejano observes in his analysis of the public meanings of climate change, there is something that appeals more powerfully to the human mind than science. That is the power of narrative. He importantly and correctly observes, "The narrative [that climate] sceptics adopt . . . may be based on a story that is not primarily about the strength of climate science or even about climate per se" (Lejano, 2019: ES419). But this argument needs applying beyond sceptical narratives only. Although contrarian interpretations of climate change are always about more than just science, so too is the process of narrative construction embarked upon by transformative radicals examined in the next chapter. As too did the reformed modernists of the previous chapter. All three interpretations of the idea of climate change offered in Chapters 3, 4, and 5 therefore lean heavily on scientific claims. At least they do so rhetorically. But in all three cases it is climate science interpreted so as to bolster a wider political narrative justifying what should or should not be done about climate change.

Notes

1 A contrarian is someone who holds a contrary position against an assumed or realised majority. It is often applied to those who challenge or reject a scientific consensus on some particular issue, particularly in cases where scientific evidence bears on political, social, or cultural controversies.

2 Palin, S. (2009) Sarah Palin on the politicization of the Copenhagen climate conference. *Washington Post*, 9 December, www.washingtonpost.com/wp-dyn/content/article/2009/12/08/AR2009120803402.html [accessed 12 August 2020].

3 But not necessarily so. The Spanish political party Podemos would be viewed by most analysts as a populist party of the political Left, as might also be – at least in the eyes of some – Emmanuel Macron's original formation in 2016 of La République en Marche!, Jeremy Corbyn's UK Labour Party between 2015 and 2020, and Bernie Sanders' two abortive bids for the US Presidency in 2016 and 2020. See Beeson (2019) for a helpful discussion of this question.

4 Mazoue, A. (2019) Le Pen's national rally goes green in bid for European election votes. *France* 24, 20 April, www.france24.com/en/20190420-le-pen-national-rally-front-environment-european-elections-france [accessed 2 July 2020].

Further Reading

Naomi Oreskes and **Eric Conway's** *Merchants of Doubt: How a Handful of Scientists Obscured the Truth on Issues from Tobacco Smoke to Global Warming* (Bloomsbury Publishing, 2010) is one of the more widely read accounts of climate scepticism, although it is heavily slanted to the American version of political conservatism. For a comparable account of contrarianism but from a UK perspective, there is **Richard Black's** *Denied: The Rise and Fall of Climate Contrarianism* (The Real Press, 2018). For a corrective to both these books, I recommend **Greg Garrard** and **colleagues'** *Climate Change Scepticism: A Transnational Ecocritical Analysis* (Bloomsbury Academic, 2019), which approaches scepticism in a geographically nuanced and more even-handed and penetrating manner. In similar vein **Raul Lejano** and **Shondel Nero** nicely illuminate the contrarian mindset in *The Power of Narrative: Climate Skepticism and the Deconstruction of Science* (Oxford University Press, 2020). This goes much deeper than Oreskes and Conway's or Black's conspiratorial accounts of scepticism. Also offering a sympathetic, yet rigorous, treatment of contrarianism – and revealing its culturally and geographically situated nature – is **Eric Swift's** *Chesapeake Requiem: A Year with the Watermen of Vanishing Tangier Island* (Dey Street Books/Harper Collins, 2018). Swift offers a superbly written and engaging account of how the crabbers and lobstermen of Tangier Island make sense of their changing climatic and coastal environment. To understand how climate science is (re-)interpreted by contrarian scientists, **Andrew Montford's** *The Hockey-Stick Illusion, Climategate and the Corruption of Science* (Stacey International, 2010) would be the place to start. In *The Climate Files: The Battle for the Truth about Global Warming* (Guardian Books, 2010), **Fred Pearce** offers a very readable journalist's account of the Climategate controversy. A more incisive and very valuable sociological critique of climate science and the scientific ideal of scepticism is offered in **Meritxell Ramírez-i-Ollé's** *Into the Woods: An Epistemography of Climate Change* (Manchester University Press, 2020). **James Painter's** *Poles Apart: The International Reporting of Climate Scepticism* (Reuters Institute for the Study of Journalism, 2011) is worth reading for his comprehensive examination of the media reporting of climate sceptic views. Finally, one of the best insights into a lukewarmist account of what climate change means for policy comes from the late conservative thinker **Roger Scruton** in *How to Think Seriously About the Planet: The Case for an Environmental Conservatism* (2012).

QUESTIONS FOR CLASS DISCUSSION OR ASSESSMENT

Q4.1: Why do value-based political and ideological arguments about responses to climate change so easily locate themselves around the nature and credibility of scientific evidence?

Q4.2: In what ways do different interpretative narratives of climate change become closely attached to people's identities? Is this process inevitable?

Q4.3: How far has the divisive partisan nature of American political life infected public attitudes to climate change in non-American cultures?

Q4.4: Would you describe yourself a sceptical contrarian in any of the senses used in this chapter? About what exactly are you "contrary"?

Q4.5: What is different and what is similar between the lukewarmist position on climate change (this chapter) and the ecomodernist position (see Chapter 3)?

5

TRANSFORMATIVE RADICALISM

. . . Climate change mobilised . . .

Introduction

In 2018 the world became aware of a 15-year-old Swedish schoolgirl called Greta Thunberg. She gained initial public attention in August 2018 by refusing to attend school and protesting against the Swedish government's climate policies. She sat outside the Parliament building every day for three weeks with a hand-written placard that simply read *Skolstrejk för klimatet* – "school strike for climate". Thunberg's protest was widely promoted by climate campaigners, social movements, and environmental media and her international public visibility grew rapidly during the following months. A wider range of mainstream social and news media began amplifying her slogans and she became valorised by many climate advocates and social movements as an iconic "youth voice" of climate protest.

Thunberg was subsequently invited to speak at the COP-24 intergovernmental climate negotiations held in Katowice, Poland, in December 2018. She started her speech with simple bluntness that became her hallmark: "My name is Greta Thunberg. I am 15 years old. I am from Sweden. I speak on behalf of Climate Justice Now". Within 12 months Thunberg became a worldwide media phenomenon. Her initial lone protest catalysed a larger school strike movement – now called Fridays for Future (FFF) – which by the spring of 2019 had spread to over 100 countries, including Australia, Chile, Estonia, India, and Turkey. If any political gathering or organisation wished to appear serious about tackling climate change – whether the World Economic Forum in Davos, the United Nations in New York, national parliaments and assemblies, protest movements such as Extinction Rebellion (XR) – during 2019 it seemed necessary to have Thunberg appear on their platform.

But what is Thunberg's interpretative narrative of climate change? Is she a political "radical" and, if so, in what sense? At one level her message appears revolutionary. She appears to have little faith in the political class as her speech in Katowice testified: "You have ignored us in the past and you will ignore us again. . . . We have come here to let you know that change is coming, whether you like it or not. The real power belongs to the people". "Change is coming, you have ignored us, power is with the people". These are the familiar slogans of revolutionaries for centuries past. And yet it is hard to discern a coherent world vision or a political programme in Thunberg's rhetoric. There is certainly no international political strategy. By attacking the political class, for example, she is indirectly shutting down discussions between the developed and less developed nations (Germani, 2019). In 2019, Thunberg's voice was new and distinctive. Her youth, simplicity, and seeming innocence were her authenticity. It was a call for popular mobilisation from an unlikely teenager – "climate change is an emergency, action is urgent, you the political class have failed me, failed us, failed my generation". And yet – as we shall see in this chapter – Thunberg's message is also strangely post-political (Mouffe, 2005). Her call to the world's leaders to "listen to the scientists" is an example of the "science-based" positions on climate change with which Chapters 3, 4, and 5 are concerned.

***

This chapter offers an account of the interpretative position on climate change I label "transformative radicalism". It is a stance characterised by two commitments: An adherence, often tenacious and passionate, to the findings of climate science and a deep dissatisfaction with the ideology and institutions of the current world order. Transformative radicalism – as with reformed modernism (Chapter 3) and sceptical contrarianism (Chapter 4) – is characterised here as a "science-based" position on climate change. It leans heavily on the adequacy of science for supplying authoritative knowledge about the physical condition of the world. In particular, climate radicals draw legitimacy for their political programme from the urgency for change believed to be warranted by scientific claims about climatic future. They emphasise the more dramatic scientific commentaries – for example about runaway climate change and tipping points – whilst chastising some climate scientists for their reticence or conservatism when making public pronouncements about the future of climate.

Radicals share with modernists and contrarians a respect for the diagnostic authority of science, rightly conducted. Science earns its social status by revealing "how the world works", climate science by showing how the climate system functions. But climate radicals make a move that puts them on a different pathway of political action. They are convinced that the scientific prognosis of the climatic future demands a major re-structuring of the world's political economy and a re-orientation of human values. They seek not *re*formation, but *trans*formation. Their dissatisfaction with the current world order extends to the globalist politics of the United Nations, the hegemony of globalisation and market neoliberalism, the adherence to twentieth-century models of conventional economic growth, the unremitted injustices of historical colonialism and – for some – to democracy itself. Climate change is a mobilising idea to inspire people to "rise up" and challenge – even to overthrow – the very institutions that the reformed modernists of Chapter 3 believe, if reformed, offer the best prospect of arresting future climate change. Advocates of radical transformation draw inspiration from various intellectual traditions, including eco-socialism, deep ecology, and post-colonialism.

The chapter draws upon the disciplines of *political ecology, transition studies,* and *critical geography* and is structured as follows. I begin by offering a brief portrait of the transformative radicalist position, rooting its origins in the political radicalism of neo-Marxism, anti-capitalism, and eco-centrism. This portrait also explains how climate science is interpreted by climate radicals and how the authority of science is appropriated for use as a political resource. I then explore in subsequent subsections two important facets of this position on climate change: Its trenchant critique of the ideology of growth and the importance of reclaiming "the local" as a site of political resistance. The chapter concludes with an overview of climate social movements. The instincts of reformed modernists lead them to work from "within the system", whereas radicals deem social movements to be the necessary and effective engine of change for bringing about the sought-for transformation of the world order.

Radical narratives

Over the years radical environmentalism has enrolled romantics and rationalists, socialists and conservatives, global northerners and southerners. A roll list of names would include John Muir, Aldo Leopold, Arne Næss, Rachel Carson, Paul Ehrlich, Vandana Shiva, Sunita Narain, and Bill McKibben. Their narrative is simple. The deteriorating condition of what was conceived as an untrammelled nature functions as an index which reveals the greed, violence, exploitation, and injustices of cultures, societies, and individuals. There are two sides to this narrative. On the one hand is the attention it draws to the abuse and destruction of nature. On the other is the focus on the rapacious power of individuals, social institutions, and technologies that wreak such havoc. Radical environmentalism seeks appeal to human reason, emotion, and imagination to activate political movements. Its narrative sweep – almost religious in its breadth and fervour – moves from environmental violation, destruction, and collapse to political resistance, revolution, and, eventually, restoration. As convincingly outlined by Lawrence Buell in *The Environmental Imagination: Thoreau, Nature Writing, and the Formation of American Culture* (1995), hope of a new and better world has always been part of this trope.

It is from this narrative vantage point that transformative radicalism interprets the new science of climate change. Climate change is the latest in the line of human violations of the natural world, different only this time in its scale of action, in its temporal and geographical reach. Since climate has been made global – as we saw in Chapter 2 – then climate *change* becomes inescapable for all. And the imprint that climate change leaves on human and non-human life bears the unmistakeable mark of the structures of power and the distribution of powerlessness that modernity has created. Climate science, correctly interpreted, becomes a powerful and essential ally in this story. Scientists are to be listened to, the truth is to be spoken to power and any traces of reticence, caution, or conservatism amongst climate scientists are to be expunged.

A good example of the extreme interpretation of climate science that can be inspired by this narrative comes from the work of Jem Bendell, professor of sustainability leadership at the University of Cumbria in England. In July 2018 Bendell released an informal and unreviewed academic paper titled "Deep Adaptation: A Map for Navigating Climate Tragedy".[1] This quickly became a rallying point for many climate radicals. It has been downloaded hundreds of thousands of times and been translated into ten other languages. Bendell's reading of the science of climate change leads him to offer his readers a despairing vision of the future. . .

> the evidence is mounting that the impacts [of climate change] will be catastrophic to our livelihoods and the societies that we live within. Our norms of behaviour, that we call our 'civilisation', may also degrade. When we contemplate this possibility, it can seem abstract . . . But when I say starvation, destruction, migration, disease and war, I mean in your own life. With the power down, soon you wouldn't have water coming out of your tap. You will depend on your neighbours for food and some warmth. You will become malnourished. You won't know whether to stay or go. You will fear being violently killed before starving to death.
>
> (Bendell, 2018: 13)

Bendell and others like him – for example David Wallace-Wells' book *The Uninhabitable Earth: Life After Warming* (2019) – offer what they claim is the "true" scientific account of the climatic future. For Bendell, in contrast to Thunberg's uncritical acceptance of the IPCC as "the trusted voice of science", the mainstream assessments of the IPCC are too conservative. Where previously he had "taken the analyses of the IPCC as authoritative", Bendell now began to see the IPCC as "very compromised" and designed to "keep people in the room rather than running for the hills" (interview quoted in Reisz, 2019). Bendell, and new social movements such as Extinction Rebellion, furthermore draw attention to a different version of climate scepticism or denial than those we looked at in Chapter 4. For them, climate denial is not denying the *reality* of climate change. It is about denying the *severity* of climate change. According to these extreme radicals, institutions like the IPCC, some climate scientists, and those governments who rely on them, are not "telling the truth" about the future that awaits. The scientific reality is much worse than such mainstream scientists speak of.

I use the example of Bendell to highlight the importance for climate radicals of grounding their political programme of social transformation in a "truthfully told" account of climate science. Whether it is Thunberg's affirmatory plea to "listen to the scientists" or Bendell's extreme claims of climate denialism within the mainstream scientific profession, climate radicals deploy science – indeed, *need* science – to establish an authoritative and horrifying account of the climatic future. It is a pessimistic and **declensionist** vision – now scientifically authenticated, they would claim[2] – which then warrants the programme of political change they advocate. For those promoting radical transformation, climate change is more than a *challenge* to modernity. It is a crisis *of* modernity. It is a planetary emergency for which business-as-usual politics, sustaining the status quo or even incremental reform are inadequate responses.

The position of transformative radicalism finds different modes of expression in different parts of the world. The Earth Strike Movement – which used the Thunberg-inspired FFF school strikes to call for a global climate strike in September 2019 – offers one articulation in its Earth Strike Petition (see Box 5.1). The oppositional politics inspired by climate radicals frequently draws together activists committed to a wide range of ecological, social, and political justice issues. New social movements seeded by resistance to the drivers of climate change – such as 350.org, Climate Justice Now, XR, the divestment movement – often enlist members committed to other radical causes, whether anti-capitalism, anti-globalisation, Indigenous peoples' rights, or anti-patriarchy. Climate change thus functions as a discursive space within which to advance an intersectional understanding of discrimination against social and political identities rooted in gender, race, class, sexuality, and ability.

BOX 5.1 THE EARTH STRIKE PETITION

"The political, economic, and social institutions that govern our global civilization have failed to react to an ever-worsening environmental issue. Institutions and individuals that presume to be our leaders should be held responsible for protecting and overseeing the ecological systems

(*Continued*)

that make human life possible. It has become apparent that the most powerful institutions worldwide are wholly incapable of living up to this most basic responsibility, and that they have been utterly negligent in heeding the warnings of the thousands of experts that study the Earth's climate and the types of environmental degradation that accompany industrial civilization. . . [they] have repeatedly warned that if we continue to consume fossil fuels at the rate and scale that we currently do worldwide, catastrophic sea level rises will occur; agricultural blight will become universal; and ecosystems will be completely destroyed . . . there are many more terrors that await the human species and all other organisms on the planet if we choose to ignore these warnings."

"The warnings and potential courses of action that might mitigate the worst of these changes have been suppressed and ignored by our most influential and powerful leaders and institutions for decades now. These institutions, though they may seem superficially dissimilar, operate primarily according to an ideological commitment to unlimited economic growth and to the extreme concentration of wealth and power in the hands of a very small group of individuals and organizations. . . . We call on you [leaders, academics, scientists, and all other professionals] to recognize the gravity of our situation, the illegitimacy of the current status quo, and come together to engineer a more intelligent mode of civilization."

"We demand that this unsustainable and suicidal system of economic and governmental policies be brought to an immediate and decisive halt. . . . We will redirect our energies from the perpetuation of our current unsustainable system of domination and destruction to the development of a more humane system that seeks justice and well-being for the many and not the few, beginning with the reclamation of long-term environmental stability and sustainability. Together, we can democratize the methods by which we organize and bring about sustainability in our time. The choice is ours, but time is running out."

Source:

Extracts from the Earth Strike Movement: www.earth-strike.com/ [Accessed 2 July 2020].

For those drawing upon Marxist inspiration for their programme of transformation, the root cause of climate change is to be found in the underlying structures of modernity: The political, social, and economic systems that perpetuate not just a fossil fuel economy but wider forms of political and social oppression. Andreas Malm's *Fossil Capital: The Rise of Steam Power and the Roots of Global Warming* (2016) offers a good account of this view. A neo-Marxian analysis would suggest that the structures of modernity are destroying the living conditions of the planet for humans and non-humans. This is what John Bellamy Foster refers to as Marx's "**metabolic rift**", a break in the cyclical processes of ecological exchange (Foster, 1999). Furthermore, the structural analysis of societies pioneered by the likes of Marx, Lenin, and Gramsci highlights the importance of "key moments" for the reordering of those societies. Power is most dangerous when it is hidden. Power needs to be exposed and resisted so that at moments of crisis, social movements of resistance and transformation can catalyse the necessary change. As one Marxian political economist explains: "Environmentalists call for socio-ecological transformation, Marx called for revolution; both mean essentially the same" (Pirgmaier, 2018: 265).

Calls for radical transformation of society in response to climate change can also be inspired by the related ideology of eco-centrism. This too finds its origins in a reaction against modernity, but there are different antecedent traditions from which this ideology has emerged. These include the European **Romantics** of the early nineteenth century, the American environmental awakening of the 1960s, or, more latterly, the political expressions of environmental solidarity inspired by Indigenous cultures and ontologies (see Chapter 6), such as those embedded in Bolivia's constitutional "Law of the Rights of Mother Earth". This deep ecological diagnosis of the problem of climate change is less focused on the social structures and oppressive power of extractivist capitalism. Rather, climate change is seen as an inevitable outcome of a human-centric rather than an eco-centric disposition towards the natural world. Environmental sociologist Eileen Crist has explored the implications of human-centrism – and its associated propensity for displaying attitudes of entitlement, supremacy, and mastery – in much of her work. For Crist, radical transformation inspired by climate change requires getting beyond human exceptionalism and dominance: "A human-centric worldview is blinding humanity to the consequences of our actions" (Crist, 2018: 1242). Conversely, an eco-centric worldview requires "scaling down" and "pulling back" from incessant human expansion.

Against growth

One of the central pillars of the transformative radicalist position is the belief that through excessive material consumption and population growth humans are destroying the natural world upon which they depend. The world has either already transgressed – or will shortly reach – important planetary boundaries with respect to ecosystem functions or resources. These boundaries define thresholds in the Earth's biological, chemical, and physical systems, which, if not exceeded, represent a "safe operating space for humanity" (Rockström et al., 2009). Beyond these boundaries, the risk grows of causing irreversible environmental changes that threaten the stability of Earth's systems and human civilisation.

The planetary boundaries framework is a new interpretation of the old "limits to growth" debate that erupted in the early 1970s (Raworth, 2018). It offers a pessimistic story of impending ecological destruction and – for some commentators – presages the possible extinction of humanity (Box 5.1). From this conviction, the carbon markets, green growth, and technological solutionism proposed by reformed modernists are an inadequate response. Arresting climate change demands a more far-reaching transformation of human societies. Some elements of this transformation are cultural and behavioural. These might include limits on consumption and diet, such as restricted meat-eating. Or restrictions on lifestyles choices, such as through *flygskam* ("flying shame"), which confronts one of the powerful symbols of modernity, air travel. Other elements are technological, such as the transition to a fully renewable energy system. But at the heart of the requisite radical transformation is a protest against the ideology of "growth" fuelled by ever-growing consumption. This is the challenge implied by Crist's "scaling down" and "pulling back".

Challenging "growth" places the programme of transformative radicalism in clear distinction to the position of reformed modernism, in which green growth is good. And it also places it in contrast to the position of sceptical contrarianism, in which *conventional* economic growth is good. For climate radicals however, climate change requires a thorough questioning of the models and metrics of economic growth, whether green or not. Transformative radicalism uses the idea of climate change to push back against one of the central myths of modernity, namely that economic growth is a measure of human progress (Wagner, 2016). Increased resource consumption, expansion of traded goods and services, sustained economic growth at 2 per cent per annum . . . all this is incompatible with a stable climate. Kevin Anderson and Alice Bows-Larkin put the claim bluntly: "Continuing with economic growth over the coming two decades is incompatible with meeting our international obligations on climate change" (Anderson & Bows-Larkin, 2013: para. 1).

The idea of climate change can inspire the growth paradigm to be challenged in a number of ways. In contrast to carbon markets and the social cost of carbon favoured by environmental economists' vision of green growth (see Chapter 3), those promoting radical transformation lean heavily on the commitments and insights of ecological and/or feminist economics. Advocacy may be for either degrowth or zero-growth (i.e., steady-state economies; see Van Den Bergh, 2017). In other words, a totally different understanding of growth and prosperity is demanded to that which has prevailed under modernity. The argument for degrowth has always found strong advocacy within the French tradition of sustainable economics (D'Alisa et al., 2014). Inspired by the "limits to growth" debate set in motion by the Club of Rome in 1972 (Meadows et al., 1972), the French social philosopher André Gorz asked half a century ago whether *décroissance* – "degrowth" – would be necessary for restoring the Earth's balance. Similarly, the Colombian scholar Arturo Escobar has long been critical of Westernised concepts of growth and development.

Degrowth requires a re-organisation of society, such that economic output as conventionally measured declines over time. It is usually interpreted as a reduction in material consumption and physical infrastructure, implying a reduction in working hours, restrictions on advertising (and hence consumption), or, more broadly, a stance that is anti-capital. Degrowth reduces pressure on the world's finite resources and alleviates pressure on the climate system by reversing the growth in greenhouse gas emissions (Figure 5.1). The ideology of degrowth aspires to construct a society that "lives better with less" (Kallis, 2011).

Advocacy for degrowth inevitably opens new discursive and political spaces for re-thinking the question of what it means to "live well". What does it mean for humans to prosper? For transformative radicals, climate change offers a window of opportunity for rethinking deeper questions of human flourishing,

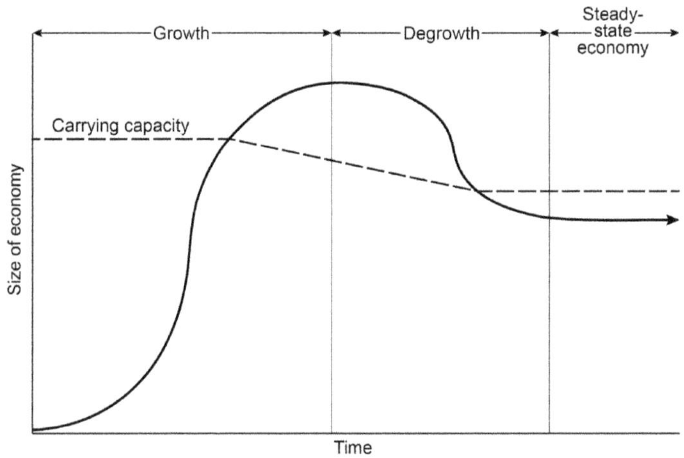

Figure 5.1 Conceptual relationship between trajectories of "growth" and the planet's carrying capacity.
(*Source*: Redrawn from O'Neill, 2012).

similar to those prompted by the SARS-CoV-2 pandemic of 2020. Future ways of living should be based on different values to those that have dominated the past. Rather than values expressed through measures of material well-being, preferred values often include the **ethics of care**, the regeneration of nature, and a fair distribution of the benefits of creativity and innovation. Human prosperity comes from enjoying the immaterial benefits of living in society – relationships, belonging, community, creativity – as much as it comes from enjoying the material benefits of "progress". In the minds of some, elements of "the economy of tomorrow" (Jackson, 2017) are already being created through the expansion of worker cooperatives, **time-banks**, and community-owned renewable energy systems. These organisational models of (usually local) economic activity create regenerative and distributive systems believed to support prosperity for all.

At this point it is worth revisiting the notion of the green new deal. Yes, there is a version of this political programme for tackling climate change that sits easily within a reformist tradition. We saw some of the characteristics of such a programme in Chapter 3. But there is also a more radical version of the green new deal which is predicated on a very different vision of economics. It approaches economic *and* social transformation jointly through the lens of intersectionality. Ann Pettifor was one of the original founders of the 2008 green new deal formulation in the UK and in *The Case for the Green New Deal* (2019) she outlines her vision. Pettifor resists the terminology of degrowth since it "simply reinforces the concept of growth". Instead, she argues for a "steady-state" economy, "an economy with a relatively stable, mildly fluctuating product of population and per capita consumption" (p. 66).

For transformative radicals however, the green new deal is about much more than economics (Hathaway, 2020). It offers a larger and more comprehensive vision than is offered by modernist reformers of "the kind of world we want". Pettifor explains this larger intersectional vision. For her, the green new deal must deliver social justice for all classes. This will entail developing a labour-intensive economy with well-paid jobs, rather than perpetuating an insecure gig economy and bonded labour. It should promote an expanded welfare safety net, if not a basic income for all. It must enshrine women's rights over their own bodies. And it must create a society with more walking and cycling and less flying and meat-eating. For climate radicals such as Pettifor – and political campaigners such as Naomi Klein and Alexandria Ocasio-Cortez – the green new deal stands in opposition to those who "consider it utopian to believe society can end a deeply entrenched system of racialized capitalism . . . that poverty, racial and gender inequality and injustice are not a result of globalised capitalism. . . [or that] capitalism's hyper-globalisation is working just fine" (Pettifor, 2019: xvi).

Localism

A second pillar for many expressions of transformative radicalism is a focus on the local. The ideology of "localism" sets itself against the imperialist forces of globalisation, which it sees as fuelling the underlying drivers of climate change. This localist vision of the 'good life' embraces forms of communalism composed of "small-scale economy[ies] based on norms of frugality and simplicity" (Karlsson, 2016: 23). As we saw earlier, conventional economic growth is not believed by transformative radicals to be essential for human prospering. International agreements on climate change – while valuable – are insufficient to deal with the underlying drivers and dynamics of global capital. Rather, it is through various forms of localism that harmony with nature can best be fostered. In this line of thinking, meaningful action on climate change must always find its primary expression in local places, even if connected across space with other localist actions.

These expressions of localism resonate with earlier traditions of environmentalism that are intensely located in place. But this sets up an uncomfortable tension for those pursuing social transformation, as Ursula Heise explores in her book *Sense of Place, Sense of Planet: the Environmental Imagination of the Global* (2008). Is the pursuit of local identity and its ties to place "an essentialist myth", asks Heise (p. 7), "or [is it] a promising site of struggle against both national and local domination?" There is an intriguing convergence of thinking here between transformative radicals and environmental conservatives with their valorisation of *heimat* (see Chapter 4).

The localist vision espoused by transformative radicals is expressed through a variety of new social movements, for example Transition Towns, eco-cities, eco-villages, eco-neighbourhoods, and **living labs**. The Transition Towns movement, for example, inspires grassroots community projects that aim to increase self-sufficiency to reduce the potential effects of peak oil, climate destruction, and economic instability (Hopkins, 2008) (Figure 5.2).

Figure 5.2 Publicity image from the Transition Towns movement.
(Source: *The Transition Handbook* by Green Books, reproduced by kind permission of the artist, Jennifer Johnson).

It started in 2006 in the UK with the small town of Totnes in Devon and has expanded in the years since then to now embrace over 400 towns and communities in over 40 countries. "Transitions" initiatives are always specific to the characteristics of a local area and culture. They are constructed as intrinsically participatory and inclusive, with a focus on community-scale demonstration projects – for example community gardening, repair shops, herbal walks, and cooperative solar energy production. Their attempt at local inclusivity has sometimes brought criticism for offering solutions to climate change that are post-political (see Box 5.2 and the work of Anneleen Kenis).

There is a close connection between localist projects of radical transformation such as these and theories of social change and **deliberative democracy**. This relationship has been carefully explored in Frank Fischer's *Climate Crisis and the Democratic Prospect: Participatory Governance in Sustainable Communities* (2017). Fischer argues that operationalising democracy at a global scale to arrest climate change – through treaties, civic mobilisation, and international cooperation – will be impossible. Such globalist ventures will always fall prey to technocratic, authoritarian, or populist impulses, not least under declarations of "planetary emergency". Instead, he advocates for a participatory and inclusive form of democratic decision making that can only ever really work at the local scale. For Fischer, it is at the local scale that a revitalised environmental democracy can best – can perhaps only – be realised. However, another plausible model for a sustainable democracy radically different from today's modes of socio-political organisation would be "green republicanism". Heidenreich (2018) shows how this form of participatory eco-democracy is already operating in European cities such as Copenhagen and Tübingen. The underlying philosophy is that citizens must be central in local decision making if they are to accept the impacts of the transformations that such collectively binding experiments in socio-technical organisation will bring about. Securing ambitious goals of city-based greenhouse gas emissions reductions and enhanced local climate resilience requires the cultivation of strong eco-localist identities. Wide-scale participation of citizens in local governance is essential to this end.

These various localist visions of socio-ecological change can be grouped together under what are sometimes called "enabling approaches" for securing radical transformation (Scoones et al., 2020). Enabling approaches seek to empower individuals and communities to take action on their own behalf. Groups and individuals are engaged at the local level to demonstrate their capacity to participate as citizens, to make their interests heard on matters of local concern. Examples of such local citizen protests would include community forest activism such as the Chhattisgarh Community Forest Rights Project in central India and resistance to new fracking developments in the UK and USA. Alliances and social networks are formed through grassroots activism, which sustain platforms of local solidarity. These platforms share the burdens and benefits of transformational political processes. It is believed possible that large-scale system transformation can be catalysed through local community-based eco-activism. And it will do so in ways that resonate with the values of embedded citizens rather than with the remote institutions of the state. Indeed, since they may rely upon the adversarial politics of citizen mobilization, these forms of citizen engagement and local eco-activism are often not welcomed by the state.

These localist visions of eco-communities, deliberative and environmental democracies, and the related enabling approaches to political activism, raise questions about their scale and efficacy. By design, localist approaches to transformative change in response to climate change privilege local identities, local concerns, and local solutions. And yet they operate in an inherently globalised world where the dynamics of finance, economics, technology, and power operate through networks and not within territories. For some sympathetic critics of climate radicalism, localist visions of change offer an overly optimistic – if not naïve – view of "how 'enabled' communities can induce needed structural changes to escape traps of poverty or oppression" (Scoones et al., 2020: 68). In this view, eco-localism may simply be a retreat from meeting the full scale of the challenges posed by climate change to a transformative view of the future. It is important therefore to look in some more detail at the wider theory and practice of social movements.

BOX 5.2 ANNELEEN KENIS (B.1980) – CLIMATE CHANGE AND THE POST-POLITICAL

Anneleen Kenis is an interdisciplinary geographer from Belgium, with a background in political/human ecology, sustainable development, and psychology. After working for environmental organisations for several years, she embarked upon her PhD at the University of Leuven, Belgium, studying processes of politicisation and depoliticisation in relation to climate change. She obtained a doctoral degree in 2015 for her dissertation "From Individual to Collective Change and Beyond: Ecological Citizenship and Politicisation".

Kenis criticises the framing of climate change by social actors in ways that occlude the real nature of political choice. In this she takes inspiration from the work of political theorists Chantal Mouffe and Jacques Rancière and critical geographer Erik Swyngedouw in their designation of "the post-political". As Kenis explains, "A discourse is post-political when it conceals power, exclusion or conflict or, more precisely, when it conceals the fact that a discourse is in itself an exercise of power" (Kenis, 2019: 845). For example, drawing upon fieldwork in Antwerp and London, Kenis showed how air pollution becomes framed as a technical issue to solve, rather than the result of a series of political choices that need challenging (Kenis, 2020). The danger for Kenis – and similar thinkers – is that post-political discourses suppress or deny the legitimacy of conflict and thereby threaten to undermine democracy.

With climate change she has pursued a similar line of critique. Kenis' complaint is that many contemporary discourses on climate change are profoundly post-political. Just because there are multiple public voices advocating for or against certain climate change policies does not guarantee that the underlying political choices are exposed. In her study of two social movements – Transition Towns and Climate Justice Action – Kenis shows how both movements fail to fully disclose what is at stake politically. Both movements fail to engage adequately in the political agonism that Mouffe and others regard as the hallmark of a democracy (also see Box 5.3). Transition Towns, by seeking a comfortable consensus around the importance of "the local" and by psychologising other political positions, denies the importance of power relations and of social inequalities. And Climate Justice Action, by intensifying the "us-them" distinction, risk placing themselves beyond the possibility of agonistic debate. Kenis argues that it is important to politicise climate change at a level necessary for democracies to discern and debate the different political visions of the future that are at stake. Her work attempts to safeguard a democratic and free society by keeping open the space for a plurality of conflicting perspectives or positions.

Read:

Kenis, A. and Mathijs, E. (2014) Climate change and post-politics: Repoliticizing the present by imagining the future? *Geoforum*. 52: 148–156; Kenis, A. (2019) Post-politics contested: Why multiple voices on climate change do not equal politicisation. *Environment and Planning C: Politics and Space*. 37(5): 831–848; Kenis, A. (2020) Science, citizens and air pollution: Constructing environmental (in)justice. In: *Toxic Truths: Environmental Justice and Citizen Science in a Post-Truth Age*. Davies, T. and Mah, A. (eds.). Manchester: Manchester University Press; See also Swyngedouw, E. (2010) Apocalypse forever? Post-political populism and the specter of climate change. *Theory, Culture and Society*. 27(2–3): 213–232.

Social movements

The idea of climate change was assimilated into the civic engagement and political activism of many established environmental NGOs towards the end of the twentieth century. Greenpeace, Friends of the Earth, and the World Wide Fund for Nature (WWF) were some of large multinational environmental NGOs that first started campaigning on climate change in the 1980s and 1990s. These NGOs were already challenging some of the political, economic, and technological systems of modernity because of their adverse environmental outcomes. To varying degrees these corporate NGOs can be seen as advocating early forms of radical transformation in response to climate change.

They were followed by a new style of social movement whose sole inspiration was resistance to the underlying drivers of climate change. The growing scientific, political, and public attention given to climate change in the first decades of the new century – attention that was itself partly a result of these earlier NGO campaigns – seeded new climate social movements, such as 350.org, Rising Tide, Plane Stupid, the Dark Mountain Project, Climate Justice Now, Transition Towns, XR, FFF. Some of these movements were born as local or national initiatives, campaigning on a narrow front. Others quickly gained international reach and visibility, while some coalesced to form transnational network initiatives, such as Alliances for Climate Action or Climate Justice Now. For these movements the idea of climate change is understood first and foremost as a catalyst for accelerating social and political change. Many mobilise around a vision of transformative radicalism.

Social movements are organised expressions of social behaviour explicitly directed towards political action. They mobilise human and material resources in ways to affect political change, usually manifesting through publicly recognized forms of civic protest or direct action. Their collective identity is based in a set of shared values and beliefs that empower those who identify with them (Jamison, 2010). Different climate social movements adopt different goals and means of achieving those goals (Table 5.1). Their informal character means that social movements usually exist as networks, in recent years using social media to assemble forms of social cohesion, identity, and action. For many advocates of transformative radicalism, social movements are recognised as an essential form of political activism.

Climate social movements always engage in some form of "communicative action" (Habermas, 1984). This may be achieved by framing climate change a certain way or through emphasizing the role of passions, emotions, and an embodied collective presence. The crowd on the street not only symbolises the lives that are threatened by what is understood to be the "slow violence" of ecological breakdown. The protest consists of those lives. Rather than having their views being represented by others – of being spoken for – social

Table 5.1 Repertoires of contention amongst climate social movements.

	Large-scale technological change	*Smaller-scale social technologies*
Policy reform, developing pre-figurative practices	NGO-led lobbying aimed at achieving policy reform (e.g. the divestment movement)	Demonstration projects (e.g. Transition Towns, low-carbon communities, climate camps)
Challenging unsustainable practices	Popular demonstrations forcing elites to act through construction of mass movements (e.g. XR, FFF)	Mass and elite direction action (e.g. Plane Stupid, Leave it in the Ground)

(*Source*: Adapted from North, 2011).

movements allow citizens to be physically present themselves. This begins to explain the powerful displays of dissent and anger that new climate movements such as XR and FFF have gained in recent years. In this sense, mass climate movements might be understood as an expression of green or environmental populism (Davies, 2019; Beeson, 2019). Mobilisations grant visibility to "the people" in bio-political form in public spaces. They become their own spectacle.

There are two dimensions fundamental to any successful social movement, defined by Jamison (2010) as the *cosmological* and the *technological*. On the one hand, a social movement is built upon ideas and ideologies – its cosmological dimension. In the case of climate change, this might be characterised by an emphasis on attending to the politics of knowledge – for example whether the scientific "truth" is being told – or by granting status to Indigenous climate knowledges (see Chapter 6). On the other hand, a social movement is built upon activities and forms of action. These would include information dissemination, social media campaigning, and practical demonstrations of both protest and constructive alternatives – a social movement's technological dimension. The direct activism espoused by many climate social movements focus on enhancing the processes and capacities of mobilisation, as much as on any substantive outcomes from their protests.

Climate social movements illustrate aspects of this theory of social change in various ways. Taking the cosmological and the technological dimensions together, social movements can be particularly important in the reconstitution of science and technology. This has been exemplified in recent years in the case of XR (see Box 5.3) and FFF. The FFF school strike movement grew rapidly during 2019 and the September 2019 global strike mobilised over six million people, mostly under 20 years of age. FFF uses forms of civic protest that have characterised many previous social movements – sit-ins, walkouts, die-ins, and strikes (Fisher, 2019). Both XR and FFF offer an effective form of public spectacle and have gained significant media exposure for this reason. On the other hand, FFF reveals some of the ambiguities and tensions in the theory of social movements identified previously. There is a disjuncture between the "success" of the protests and any ensuing meaningful political action. Yes, media attention is gained through spectacle. But it is hard to point to any meaningful change in policies, technologies, or emissions trajectories that have resulted from FFF. As Dana Fisher suggests, "one might recognize that the importance of these days of action is the action itself" (Fisher, 2019: 430).

For many climate social movements, their pursuit of a radical transformation of society as a response to climate change suggests the need for a larger narrative. As we saw earlier in the case of radical versions of the green new deal, climate change offers a flexible and powerful idea around which those pursuing an intersectional political agenda that is explicitly transformative can unite (Hathaway, 2020). For example, XR emerged from earlier social movements energised by the oppositional politics of anti-capitalism, anti-nuclear, and anti-patriarchy (Box 5.3). There is also the intersectionality of climate change, post-colonialism, and calls for reparative justice (see Chapter 6). Emancipatory decolonisation movements in the Global South have long campaigned against oppressive political regimes. For example, *La Vía Campesina* ("the peasants' way") campaigns to defend farmer's seeds, stop violence against women, recognise the rights of peasants, and for agrarian reform. Similarly, new social movements like Climate Justice Now (CJN) seek to build coalitions among different campaigning groups working towards social justice. As a "movement of movements" CJN seeks to forge a more effective challenge to power structures that perpetuate the connected evils of violence, poverty, oppression, and climate change. These postcolonial politics of climate change have been explored in Sanjay Chatuervedi and Timothy Doyle's book *Climate Terror: A Critical Geopolitics of Climate Change* (2015).

Social movements may be criticised for engaging in mainly symbolic protests that do not effectively challenge the ultimate targets of dissent. Individual and community-based social movements may make it seem like something is being done, that change is being enacted. In reality, however, some climate social movements may offer therapeutic processes of participation for angry citizens (North, 2011). They may not realise the limits of their power to change society in the face of local and global systems of domination. It is hard for citizens – often excluded from positions of political or economic power – to transform complex economic-industrial systems merely through the power of example, argument, or protest. There is also criticism of the appropriateness of social movements spawned in the Global North to campaign against the injustices of climate change that are manifest in the Global South. This raises questions about the geographies of international, ethnic, and social solidarity. This was a criticism made of the XR protestors in London in 2019, prompted by their non-inclusive social and ethnic profile. Radical social transformation as a response to climate change can mean very different things to citizens located in different political and cultural regimes.

BOX 5.3 EXTINCTION REBELLION (XR)

Extinction Rebellion gained visibility during the northern autumn of 2018, initially in the UK and then in selected other countries around the world, such as Australia, Denmark, Germany, and the USA. XR implements a very simple – and psychologically appealing – strategy for engaging people on climate change. First is an appeal to the emotion of fear. Claims such as "if you are young, you may not live to be old because of climate change" or "6 billion people will die because of climate change" are rooted in a polemical reading of climate science. The argument then proceeds that nothing is being done about this imminent catastrophe. NGO campaigning has failed, parliamentary democracy has failed, governments have failed, all fatally corrupted by fossil fuel neoliberalism. The final move is to tell people that the only way to bring about the radical transformation required is through mass civil resistance. The "civil resistance model" espoused by XR is intended to achieve mass protest accompanied by law-breaking, leading eventually to the breakdown of democracy and the state. XR claims inspiration from the Suffragettes, the Occupy Movement, and the Civil Rights Movement.

Two of the central founding figures of XR – Roger Hallam and Gail Bradbrook – had previously founded in 2016 a social movement called Rising Up! This social movement was formed of activists experienced in protest movements such as Earth First!, Plane Stupid, Radical Think Tank, and Reclaim the Power. These were all rooted in the political extremism of anarchism, eco-socialism, and radical anti-capitalist environmentalism. The founders of XR therefore had a long-held belief that the only adequate response to climate change is overturning the social order and the capitalist economic system. The real enemy of a stable and benign climate is "racialised capitalism", its fetishizing of economic growth and the centralisation of wealth and power that capitalism fuels. XR's supporters claim that in organising to protect the biosphere, the movement is working to protect the interests of all. Since "we cannot live on a dead planet", the basic environmental conditions that allow humanity to survive and flourish must be protected (Mair & Steinberger, 2019). XR has received endorsements from thousands of scientists and has inspired calls to scientists at large to embrace climate activism alongside XR (Gardner & Wordley, 2019).

XR's political strategy with respect to climate change can be positioned in the context of the political theory of agonism (Mouffe, 2005). Both agonism and antagonism are defined by political conflict, but Mouffe explains that they represent two quite different logics of political engagement. Political *agonism* frames one's opponent as an adversary, worthy of respect and debate as one seeks to craft possible lines of action. *Antagonism* on the other hand, sees one's opponent as "the enemy". One fights against them to secure victory, but with them there can be no negotiation. This would appear to be the position adopted by XR. Chantal Mouffe – and other political theorists such as William Connolly and Ernesto Laclau – advocate for political agonism over antagonism. The former is, for them, the essential condition of a democracy, the latter a dangerous condition for any democracy to fall into.

Read:

Farrell, C., Green, A., Knights, S. and Skeaping, W. (eds.) (2019) *This is Not a Drill: An Extinction Rebellion Handbook*. London: Random House/Penguin. For contrast read: Wilson, T. and Walton, R. (2019) *Extremism Rebellion: A Review of Ideology and Tactics*. London: Policy Exchange.

Chapter summary

The interpretative positions on climate change explored in this chapter seek a radical transformation of society. It is claimed that the risks and dangers of climate change demand nothing less. As with reformed modernism and sceptical contrarianism, what I call here "transformative radicalism" takes its cue from its own distinctive interpretation of scientific accounts of the present and future condition of the world's climate. This interpretation suggests that climate change threatens imminently the habitability of the planet and the possibility of a dignified life for all. To meet this challenge, climate radicals seek to *trans*form the institutions of modernity – and the ideologies upon which they rest – into new configurations, rather than merely to *re*form them. And in the case of transformative radicalism, it is clear how the ideological meaning of climate change is actively constructed. Climate change is not simply a physical threat or environmental problem revealed by science. The idea of climate change requires geographical, historical, political, ethical, and social modes of interrogation and analysis – facilitated by social science and humanities disciplines – to understand the dynamical and contested nature of what is at stake.

The chapter has emphasised two central elements to the transformative radicalist position. These are the redefinition of growth and the promotion of locally based forms of organisation to pioneer and bring about social change. Social movements are an integral part of how this vision is to be secured. Social movements open up spaces in public life for new forms of knowledge-making and socio-cultural learning to challenge existing structures of power and dominant cultural norms. This challenge to incumbent power may find expression in different ways, exemplified in this chapter by the examples of the Transition Towns movement and Extinction Rebellion. Both of these social movements are committed to using the idea of climate change to bring about transformative change in society. They both represent dynamic processes of political protest and advocacy that mobilize human, material, and cultural resources through networks that are bound together by shared values in pursuit of a common cause. Yet their respective visions of transformative change, modes of political action, and desired outcomes are very different.

***

Each of the previous three chapters has outlined a different broad position on what the idea of climate change signifies for political action in the world. In articulating these three positions I have used the idea of "modernity" as a heuristic. Thus, simply put, sceptical contrarians may be seen as affirming "Modernity 1.0", reformed modernisers as proposing "Modernity 2.0", and transformative radicals as being "Anti-Modern" or "Beyond-Modern". In each case, the science of climate change is seriously engaged with; there is great political necessity in doing so. Although not blind to their respective ideological commitments, the three positions each find political and cultural advantage in justifying their framework for political action by leaning heavily on their particular interpretation of climate science. It is for this reason these chapters have been grouped together as articulating "science-based" positions.

But what is extracted from the sprawling body of scientific evidence about climate change is different in each case. Climate reformers, contrarians, and radicals each mobilise very different sets of axioms, values, and justifications to interpret scientific evidence in ways that build distinct evidential arguments in support of their preferred political programmes. We may think of these assemblages of cognitive and normative commitments – metaphysical, moral, epistemic, religious, political – as "comprehensive doctrines"

(Rawls, 2005). Comprehensive doctrines are intelligible views of the world that offer persuasive guidance to answering age-old questions about how people should live. Although frequently deferring rhetorically to climate science, the positions outlined in the preceding three chapters are always about more than science. Which leads us now to consider other ways of constructing comprehensive doctrines that are somewhat less deferential to science.

Notes

1 A revised version of this paper was released by Bendell in July 2020, a response to criticisms he had received from other climate radicals of his "pseudoscience". Available here: http://lifeworth.com/deepadaptation.pdf [accessed 26 August 2020].

2 There are many scientists who would themselves back such a vision, for example the 11,258 signatures to the "open letter" in Ripple et al. (2020).

Further Reading

For a standard account of climate change seen through the lens of transformative radicalism, read **Naomi Klein's** influential *This Changes Everything: Capitalism vs the Climate* (Simon & Schuster, 2014). And for a more historical analysis of this position then **Andreas Malm's** *Fossil Capital: The Rise of Steam Power and the Roots of Global Warming* (Verso, 2016) is to be recommended. For a transformative radicalist view of climate change from the Global South then **Sanjay Chaturvedi** and **Timothy Doyle** have written *Climate Terror: A Critical Geopolitics of Climate Change* (Palgrave Macmillan, 2015), which also engages with the subaltern voices explored in Chapter 6. More recently **Klein** has revisited her case using the framework of the green new deal in *On Fire: The (Burning) Case for a Green New Deal* (Simon & Schuster, 2019), while **Ann Pettifor's** *The Case for the Green New Deal* (Verso, 2019) offers a comparable perspective from the UK. **Tim Jackson's** *Prosperity Without Growth: Foundations for the Economy of Tomorrow* (2nd edition, Routledge, 2017) offers the standard argument of why economic growth does not equate to human well-being. To gain a flavour of one of the new social movements espousing radical transformation read the essays in **Extinction Rebellion's** edited handbook *This is Not a Drill: An Extinction Rebellion Handbook* (Random House/Penguin, 2019). **Frank Fischer's** *Climate Crisis and the Democratic Prospect: Participatory Governance in Sustainable Communities* (Oxford University Press, 2017) is a good place to start – even if an optimistic one – when thinking about forms of democracy that are commensurate with the vision of social transformation. In *Reframing Climate Change: Constructing Ecological Geopolitics* (Routledge, 2015), geographers **Shannon O'Lear** and **Simon Dalby** have edited an excellent collection of 12 essays that critically interrogate the radical's slogan "system change, not climate change". More generally, in *The Global Warming Reader: A Century of Writing About Climate Change* (Penguin Books, 2012), the veteran eco-activist **Bill McKibben** has compiled a useful collection of writings from (mostly) the more radical or challenging voices on climate change from the past 100 years.

QUESTIONS FOR CLASS DISCUSSION OR ASSESSMENT

Q5.1: If it really *is* "capitalism versus the climate" (Klein, 2014), then what alternative forms of political economy should replace capitalism if climate is to be saved?

Q5.2: Compare and contrast the goals and tactics of the Occupy Movement and Extinction Rebellion.

Q5.3: "[Social] change comes from the edges – but it doesn't come from *staying* at the edges, guarding our countercultural purity" (Dark Mountain Project – https://dark-mountain.net/). How important is this vision of change for securing radical transformation of societies?

Q5.4: Who might benefit from re-defining growth in terms of relational well-being rather than as measured by the extent of exchange of goods and services? Who might lose out?

Q5.5: How might claims of "climate emergency" be received in less liberal political cultures such as China, Iran, or Indonesia? What might it take for the Chinese Communist Party to declare a "climate emergency"? Would this be a good thing?

6

SUBALTERN VOICES

. . . Climate change supplanted . . .

Introduction

Leboi Ole Netanga is a Maasai pastoralist who lives in Terrat, a small rural village in northern Tanzania.[1] Mr Leboi herds cattle, as have his ancestors over many generations. In November 2011 Leboi was asked to testify at a national public hearing in Dar es Salaam about the climatic challenges that the people of his village were facing. It was one of a number of such hearings taking place across Africa in the weeks leading up to the annual international climate negotiations (COP-17) to be held in Durban in early December. The hearing included testimonies from scientists, government officials, and NGO staff – and also from highland, lowland, and seaweed farmers; fishers; pastoralists; and a faith healer. Leboi opened his testimony thus:

> My name is Leboi Ole Netanga from Simanjiro district. I come from Terrat village. When we used to talk with our fellows in the village about climate change we realized that it started to occur in 1987, when drought hit the zone of Simanjiro. We are pastoralists, we tried to cultivate but we failed due to the climate change, then variation in season and no rainfall. . . . So we tried to cultivate but a few of us got maize, we tried more but we did not succeed . . . and we realized that we are in an area that does not support agriculture. So we decided to continue livestock keeping.

The rainfall on the Simanjiro plains is inherently variable. It was no surprise therefore that for Leboi speaking about the (changing) climate was the same thing as speaking about (variable) rainfall. But he and his fellow Maasai pastoralists were not sure what had caused this situation:

> Looking at history you say that the destruction of the environment causes this situation, but we are trying to find out how the destruction

> of the environment has caused it. It cannot be demonstrated clearly! We are trying to protect our environment so that it can rain, because we don't know either if it is the destruction of the environment itself or is it because of lack of rainfall. We don't know yet! We are trying more and more, but we fail. In recent years you can find there is no rainfall even for five years continuously. Nowadays there are some important pastures disappearing because of lack of rainfall.

Then something unexpected happened. Leboi seized the opportunity to put his own question to the member of the Tanzanian parliament attending the hearing:

> What is another issue? Let me ask the people from the government something: why is it that the wild animals, I mean from Tarangire [National Park], they come and eat our grass, but our cattle are not allowed to go to Tarangire to eat the grass there?

This question seemed unrelated to climate change. Yet it touched on a sensitive and important issue for the Maasai, namely the relationship between wildlife conservation and the pastoralists' cattle. Leboi's question redirected the hearing about climate change into a heated debate about the political processes that have shaped the socio-cultural and environmental landscape of which the Maasai pastoralists form a part. The various parties at the hearing kept referring back to Leboi's intervention. He had effectively reframed the debate. Rather than debating climate change on the terms set by powerful actors, Leboi focused the issue of climate change on a problem that mattered to him and his people: The longstanding problematic relationship between Maasai pastoralists and the Tanzanian state.

This example from Tanzania prompts an important question about the idea of climate change. It is a question concerning what anthropologists understand as the problem of "translation". What happens when a scientifically constructed transnational idea such as climate change enters into local cultural and political settings? How does the idea of climate change "travel" around the world and does it travel intact? What other knowledge regimes – beyond that of science – or what other forms of cultural life and political resistance does climate change encounter as it travels? Leboi was not content to talk about climate change on the terms set by the national hearing. Nor on those framed by the international discourse of climate change that would dominate COP-17 a few weeks later. For him, talk of "climate change" brought to the fore longstanding political struggles about land that had little to do with global energy, greenhouse gases, and carbon markets. Leboi wrestled ownership of the idea of climate change away from scientists and international negotiators and reframed it around concerns that were important for him and his fellow pastoralists.

***

This chapter explores how subaltern voices, such as Leboi Ole Netanga's, understand and speak about climate change. It starts from the premise that communal, personal, and ethical knowledge – knowledge that, collectively, guides how people come to be, to know, and to act in the world – exceeds the forms of climatic knowledge offered by science alone. This perspective subverts the idea of climate change as it has been constructed in – and owned by – the centres of epistemic and political power. It dismantles the assumption that climate science is sufficient for understanding the significance and implications of climate change for the peoples of the world. It therefore undermines the belief that science forms an adequate basis for political action in the world.

People make sense of their **lifeworlds** – their subjective experience of the world – by bringing together in unique combinations their ethnic identities, cultural myths, life histories, and social networks. These communal knowledges are often rooted in places, social and spiritual networks, and in political histories that are not recognised by climate science. These communal, personal, and ethical knowledges of climate change rarely fit neatly with pronouncements about climate change emanating from scientific bodies like the IPCC, international NGOs such as Greenpeace, or international institutions such as the World Bank. What the idea of climate change means to people varies geographically. Climate change looks and feels differently in different places and becomes an idea used to different ends.

To capture something of this diversity of climatic knowledges, this chapter is titled "Subaltern Voices". The descriptor "subaltern" is widely used in postcolonial studies to designate colonised peoples who are socially, politically, and geographically excluded from the hierarchies of power of a colony or from an empire's metropolitan homeland. I follow the Italian political theorist and Marxian philosopher Antonio Gramsci in using the term a little more widely than this. Gramsci used "subaltern" to define social groups that are excluded from the dominant socio-economic institutions of a society and so denied political voice. By using subaltern in this way, I bring forward voices from a variety of marginal or subordinate perspectives. These include, for example, Tlinglit hunters from Alaska and female farmers in Malawi but also voices from "marginal" communities within the Global North, such as crabbers in Chesapeake Bay and farmers in Cornwall, England.

The chapter is structured as follows. I start by revisiting some of the cultural and scientific histories of climate change introduced in Chapters 1 and 2.

This is to understand how the scientific conception of climate change travels around the world and how it is received and translated in different places and within different cultures. The following three subsections of the chapter then explore in turn some of the ontologies, temporalities, and resistances that subalterns foreground when engaging with the idea of climate change. Taken together, these perspectives establish a stronger geographical basis for understanding the idea of climate change in the contemporary world. The chapter draws inspiration from the disciplines of *cultural anthropology*, *development studies*, and *postcolonial geographies*.

How climate change travels

Chapter 1 established that the idea of climate emerges at the intersection of weather, culture, and place. Remove any one of this triad from one's understanding of climate and the complexity of what climate means for people will be missed. Climatic knowledges are strongly place-based. This turns out to be as true of scientific knowledge about climate change as it is about its vernacular equivalents. As shown in Chapter 2, scientific activity always occurs *in* places. And yet knowledge rarely stays put. It moves between people and travels between places. But not in the same way. One might argue that the great success of science is its ability to convert local facts into global facts, to ensure that facts made in "truth-spots" do indeed become facts that have universal reach. In contrast, vernacular knowledges would seem to retain stronger attachments to their places of making. Compared to scientific knowledge, they struggle to travel freely – or to gain authority easily – outside their places of origin. But even scientific facts do not *all* move freely around the world. And neither do they complete their movement without moderation or mutation, without their encountering "friction" as Anna Tsing (2005) describes it.

Scientific knowledge of climate change moves unevenly around the world through material networks, embodied agents, news and social media, and encultured discourses and artefacts. Each of these modes of travel introduces frictions which slow or modify travelling knowledge. Economic historians Peter Howlett and Mary Morgan reveal some of these processes at work in their edited collection of essays *How Well Do 'Facts' Travel? The Dissemination of Reliable Knowledge* (2010). Some facts travel further, more quickly, and with greater integrity than others. And as knowledge moves around the world it brings about unpredictable political, social, and cultural effects that reshape scientific knowledges in subtle or not so subtle ways. Some facts change as they travel; others remain intact. As Howlett and Morgan make clear, how different facts enter into public circulation – and how they gain or lose cultural and political authority as a result – requires careful study.

This thinking about circulation, translation, and resistance can be applied to the idea of climate change. As we saw in Chapter 2, over the past 50 years Western science has established a powerful understanding of climate change in which humans appear as the main agents of change. This scientific account of climate change is rooted in what I call "global kinds of knowledge" (Hulme, 2010). Over the past 25 years, in many cultural settings these kinds of knowledge have shaped public understanding of climate change. One might therefore say that this account of climate change has moved successfully "to the ends of the Earth". Yet this movement has certainly not been without its frictions. The scientific account of climate change – whilst certainly pre-eminent and commanding – has gained neither universal assent nor unimpeachable cultural authority. Not only has it been "received" very differently by different people and cultures, but active processes of translation and resistance have been at work, especially among subalterns. As historical geographer David Livingstone makes clear, "*where* scientific texts are read has an important bearing on *how* they are read. This realisation points to a fundamental instability in scientific meaning and to the crucial significance of what might be called located hermeneutics" (Livingstone, 2005: 391, emphasis added). Very different meanings than those promoted by its scientific producers and disseminators quickly become attached to the idea of climate change.

We saw this in the case of Leboi from Tanzania. Leboi resisted the scientific idea of climate change as being about long-term change in the climatic conditions of northern Tanzania brought about by greenhouse gas emissions. He reframed – "translated" – climate change as revealing a long-term struggle about how the ancestral lands of the Simanjiro plains were to be managed and by whom. This process of translation is at work in many other settings around the world. To understand how a globalised concept of climate change that science has put into worldwide circulation is remade locally, it is necessary to study what happens "when the impersonal, apolitical and universal imaginary of climate change projected by science comes into conflict with the subjective, situated and normative" (Jasanoff, 2010: 233). This task of understanding both the mobility *and* the mutability of the idea of climate change can be completed by examining four sets of contextual factors that condition the "travelling" idea of climate change: Networks of mobilisation, the mediation of social actors, the contingencies of history and culture, and sites of political resistance.

It takes substantial and sustained effort for the scientific idea of climate change to travel from the places of its production and establish itself in multiple political, social, and local settings around the world. This mobility is partly a function of the *institutional networks* within which climate science is conducted and co-opted. The IPCC has been crucial in this regard. But also important has been the patronage of numerous civic and political actors mobilised through the expanding material, discursive, and political reach of the UNFCCC. Over the past quarter century, the scientific account of climate change has become mainstreamed in public discourse through the accreditation of many national and transnational institutions and social movements. Many countries have adopted national climate change strategies, plans, and programmes into which is inscribed the scientific explanation of climate change. To take just one example from many, the Thailand Climate Change Master Plan – which guides government investment planning in that country for the period 2012 to 2050 – is formed around the scientific understanding of climate change. This too is the case for many international development institutions such as the World Bank and United Nations Development Programme.

Careful study of these networks of mobilisation, however, reveals some of the frictions and translations that are at work as the idea of climate change travels, Livingstone's "local hermeneutics". One such study from Oceania revealed how the standardised scientific account of climate change did not travel smoothly between different Pacific islands (Webber, 2015). World Bank–sponsored climate adaptation projects in these islands were designed by the bank's networks of international experts and consultants. But these project designs did not travel well from island to island, even though grounded in the same scientific knowledge of climate change. What made sense as an adaptation planning framework for the low-lying atolls of Kiribati did not "land" well in the mountainous settings of the Solomon Islands.

A second set of contextual factors that conditions the mobility and translation of the idea of climate change points to the *social actors* (or mediators) who facilitate this movement. Climate communication scholar Max Boykoff (2011) asked the question "Who speaks for the climate?" in his eponymous book dealing with media coverage of climate change. The answer is that numerous social and political actors become mediators of climate change as the idea travels around the world. These include development aid agencies like World Vision or Save the Children Fund, environmental NGOs such as WWF, journalists and filmmakers, religious charities such as CAFOD, and climate or carbon consultancies such as ClimateCare. As these actors "convey" the idea of climate change around the world through their encounters with different audiences, clients, and social formations, so it takes on different forms. Deliberately or not, climate change becomes moulded by these mediators' interpretative narratives.

A good example of this is revealed in Carol Farbotko's study of how the *Sydney Morning Herald* newspaper characterised climate change for the Tuvalu islands in Oceania (Farbotko, 2005; see also Shea et al., 2020). The scientific account of climate change rooted in global kinds of knowledge – in particular in projections of future global sea-level – is narrated by this mediator as a tragedy for the islands. Tuvaluans are represented by the *Sydney Morning*

Herald primarily as helpless victims of climate change, without agency or local knowledge, compounded by the political apathy of Australian and other Western leaders. The physical vulnerability of these atolls becomes synonymous with the islanders' social and cultural vulnerability. Climate change is told by outsiders as a story of looming and dramatic physical disaster, rather than as one focused on the knowledge, vitality, and agency of Tuvaluan society. The scientific idea of climate change "arrives" in such islands – and often in other similar subaltern locations around the world – already formed as a narrative of destruction, loss, and victimhood (Kempf, 2015). Tuvaluans are actively disempowered in this narrative and, not surprisingly, it meets local forms of resistance (see later).

As the idea of climate change moves into subaltern settings, a third set of contingencies that are *cultural and historical* condition what Jasanoff describes as the conflict between the globalised imaginary and local subjectivities. Here, there is much to learn from reception studies of the idea of climate change that have been undertaken by anthropologists and sociologists. As climate change enters and circulates in different cultures, scientific accounts of the phenomenon are reinterpreted according to local vernacular cosmologies and political subjectivities. Cultural identities and histories of place act as powerful filters for assimilating the travelling idea of climate change into local meanings. Earl Swift's ethnography of a small township community of crabbers and lobstermen on Tangier Island in Chesapeake Bay, Virginia, offers one rich example (Swift, 2018). These islanders have long experience of the fluctuations of sea, tide, and storm and have a rich set of cultural repertoires for living with such precarity. Religious faith and practice are knowledges that form an important part of this repertoire (see Chapter 8) and Tangiermen see themselves as protected through prayer. Fiercely independent, they read the signs of environmental change through their everyday theology and strong place attachment, rather than through outsiders' (scientific) explanations of why the seas are rising and the land eroding. Subalterns translate climate change in a multitude of ways as the idea interacts with different kinds of knowledge and experience, cultural histories, and competing claims to authority and expertise.

Finally, understanding the frictions generated by the travelling idea of climate change requires us to consider the different *sites and motives of political resistance*. Subalterns frequently resist scientific modernity imported from the narratives of climate change imposed from above or from outside (Ghosh et al., 2018). As with Leboi's pastoralists or the crabbers of Tangier Island, local farmers, fishers, and foresters often understand their local climates better than global scientists or state officials. They may well be suspicious of state or other outside agencies appearing to use the disciplining power of climate science to coerce them into changing tried-and-tested customary livelihood practices. Subalterns may also seek to reposition the scientific narrative of climate change in the context either of state oppression or histories of colonialism and land appropriation (Cameron, 2012). Understanding political resistance to the travelling idea of a universalised and scientised climate requires entering into the lives and ongoing efforts of subalterns to assemble alliances of different, incoherent, and hesitant actors against incumbent powers. The multiple forms of local political action catalysed by the globalised discourse of climate change cannot be controlled by a singular prescriptive reading of climate science.

Some of these underlying motives for resistance are revealed in Illia Gallo's study of how globalised discourses of climate change materialised "on the ground" amongst female farmers in Malawi (Gallo, 2018). Gallo listened carefully to how these women related climate change to their farming practices and political struggles for recognition. The travelling transnational idea of climate change – what she referred to as the "all-encompassing climate change epistemology" (p. 24) of international agencies – had little to say about the underlying vulnerabilities to weather and climate extremes these female labourers encountered. Loveness Kapininga, a young female farmer from the village of Kasache, on the shore of Lake Malawi, put it thus:

> I think I heard about climate change from organisations and radio programmes . . . I'm trying to adapt to climate change in my farming activities. . . . We just try to adapt on our own. Those that come to the community say that there is climate change and they ask how they can help. Farmers tell them about the problem and the challenges, but there is no solution. The organisations don't provide much support. I have expectations from them: I'd be happy if they provided me with information on how climate change and business are related. I'd like to receive some capital and tools to be more resilient in the face of hunger.
>
> (ibid p. 192)

Such subaltern narratives "speak back" against the scientised account of climate change, disclosing the underlying political, social, and economic conditions under which they experience their lives. Climate change becomes about far more than global carbon accounting or predicting future weather.

These examples show how the universalised scientific account of climate change is refracted – how it splinters – as it travels the world and encounters people with different cultural and political histories. Climate change *singular* becomes climate change *multiple*. Rather than being spoken in one voice – through the scientific measurements, models, and the vocabularies of global kinds of knowledge – climate change's narration necessarily becomes multivocal. Climate change is not an idea that can be transplanted from one place to another and end up meaning the same thing. Similarly, climate change does not move smoothly between scales. What climate change means locally is not simply the result of downscaling global kinds of knowledge; a numerical process of rescaling information as might be undertaken by climate modellers. The idea of climate change is supplanted as it encounters the historical, cultural, and political realities of subaltern communities. Climate change localised is never a scaled-down replica of the idea of climate change globalised.

Ontologies

In their study of what "nature" means in different cultures, Luca Coscieme and colleagues compared 63 linguistic conceptualisations of the idea (Coscieme et al., 2020). They grouped these into three meta-categories, designated as "nature integrated", "nature separated", and "nature deified". Different cultures thus reveal their different understandings of nature through language, whether or not nature is separate from the human world, nature's relationship with gods or spirits, and so on. Coscieme and colleagues argue that recognising such different worldviews creates a valuable resource when seeking to govern human–nature relations for creating sustainable futures.

In her book *Tears of Wangi: Experiments Across Worlds* (2017), the New Zealand anthropologist Anne Salmond similarly draws attention to the value of recognising what she calls cosmological – or "cosmo" – diversity. As with the term biodiversity, which points to multiple life forms, for Salmond cosmo-diversity points to the co-existence of multiple imaginative worlds. (From a Latin American perspective this phenomenon would be referred to as a "pluriverse"). Salmond's archetypal example of cosmo-diversity comes from Aotearoa New Zealand where Māori and Western cosmologies have repeatedly clashed over the past two and a half centuries. These collisions between Māori and Europeans worldviews have occurred over different understandings of what is real and what is commonsensical, about what is good and what is right. In *Tears of Wangi*, Salmond applies her cosmological analysis to waterways, land, the sea, and people, but she might equally have applied it to the question of climate change. What may self-evidently be real about climate to Western scientists is not so for the Māori – and vice versa. Different assumed realities lead to a clash of incommensurable worlds.

Learning from subaltern voices is not just a question of deciding *how* one studies climate change. This would be a question of epistemology; i.e., the methods by which one creates valid knowledge. Subaltern voices also challenge assumed realities about what sort of thing climate change is and about which valid knowledge can be made, whichever method is used. This is a question of ontology; i.e., what objects or beings actually exist. Many subalterns would appear to hold different ontologies of climate to those of scientists or rational materialists. As seen earlier – and also in Chapter 1 – the idea of "climate" as found in different cultures eludes one single definition or mode of representation. Scientific representations of climate change cannot do full justice to the different perspectives on the nature of reality that can be accommodated among a diverse humanity.

Challenging a conclusion as it is, this observation would seem to question the adequacy of a position called "ontological realism"; i.e., that there is only one objective reality recognised by all. And it is an observation that also has implications for the idea of "epistemic pluralism"; i.e., a commitment to accepting different ways of knowing an object or a being. At one level, epistemic pluralism may well be a desirable position to hold with regard to climate change. It recognises – and welcomes – that scientists, social scientists, humanists, and subalterns collect and interpret different kinds of empirical data that make climate change known. So, for the epistemic pluralist, temperature measurements, perceptions of climatic risk revealed by **Likert scales**, ethical judgements that define the social cost of carbon, and embodied life histories are all valid empirical data that reveal something about climate change. The empirical data of scientists and subalterns alike are revealing aspects of the same reality of climate change — but from different perspectives and using different methods.

But some scholars highlight the dangers of an epistemic pluralist approach to studying climate change if the analyst assumes ontological realism (Nightingale al., 2020). Without a recognition of ontological diversity, the epistemic pluralist – these critics would say – may be tempted to try to integrate these different knowledges into a single coherent account of climate change. Or, if not falling to this temptation, then for the ontological realist the value of subaltern knowledge may end up being limited solely to its ability to validate scientific knowledge of climate change (for an example of this move see Weatherhead et al., 2010). Salmond's recognition of cosmo-diversity of course challenges the ontological realist position. It is *not* possible, she would argue, to distil Indigenous or subaltern knowledges of climate change such that they can be integrated into a unified climate science. Even less can they validate scientific knowledge. And it is *not* the case, Salmond and others would say, that a singular understanding of climate change can be secured through the process of "co-production" (Bremer & Meisch, 2017), a bringing together of all stake- and knowledge-holders to craft a harmonious account of a phenomenon. Whilst some international scientific, policy, or development organisations hold out such ambition for co-production, it is challenged by scholars and subalterns alike (see Box 6.2 later in the chapter and the work of Nicole Klenk).

The conclusion from this discussion is that subaltern voices on climate change need listening to on their own terms. Subalterns train us to approach the idea of climate change with a sensitivity to (at least) the *possibility* of different ontologies. If science is de-centred from accounts of climate change – as is the case for many subalterns – then different possibilities open up for identifying the underlying causes, challenges, responses, and solutions to climate change. Resisting the assumption, instinctively made by scientists and others, that climate change is all about molecules of carbon dioxide, global carbon budgets, modelled predictions of future climate impacts, or even about local weather extremes, makes it possible to supplant the idea of climate change using very different assumptions. For example, many subaltern cultures do not hold a de-socialised view of nature in which the natural and the social are separate categories of being. In the ontologies of Māori, Inuit, Nepalese Buddhists, or First Nation Americans, for example, the referent "climate change" is always about more than a change in the physical conditions of the atmosphere, ocean, ice, or land. It would make no sense in such cultures, for example, to conceive of repairing or restoring climate by intervening solely in what science would identify as physical processes whilst ignoring social relations.

An "ontology-first" approach to understanding the idea of climate change challenges universalist accounts of climate change derived from science or other singular perspectives. It seeks to understand what climate change means by being attentive to the worldviews and political struggles of different peoples and cultures. This will often foreground ontologies that are relational rather than mechanical. In such worldviews, reality is created through the interactions of relational beings – including mythical beings embodied in mountains, forests, rivers, and so on (see Box 6.1) – rather than solely through the interactions of physical particles. This has profound implications for understanding what sort of problem climate change is and what sorts of changes might begin to constitute a solution. In a relational ontology, the causes of climate change are to be found in disturbed relationships between people, animals, the land, and the gods. Attempts to arrest climate change are therefore about restoring such relationships far more than they are about altering physical processes or substituting one technology for another.

Examples of this relational approach to restoring damaged ecosystems are readily found in various cultures. Julie Cruikshank's classic study of the Tlingit of southern Alaska (Cruikshank, 2001) showed the importance of oral traditions – and their accounts of human agency – for making sense of people's relationship with glaciers and their various movements. For the Tlinglit, the human subject is thoroughly embedded in its physical and social worlds. People are understood as subjects who take personal and collective responsibility for their behaviours. These behaviours have consequences for social and environmental phenomena. Similarly, in Inuit cosmology the idea of *Sila* connects the spiritual and material worlds within which humans are situated (Todd, 2016). *Sila* encompasses the idea of breath – the energy that drives life – but also manifests in tangible weather phenomena such as a breeze, a storm, or blizzard. "*Sila* is a life force that can be felt as air, seen as the sky, and lived as breath" (ibid: 5). *Sila* is bound up with life, with climate, with being and knowing.

The value of an ontology-first approach to the idea of climate change is that it unsettles the presumption that epistemologies rooted in Western science have exhausted the meaning of climate change. It accommodates a much broader set of perspectives rooted in different everyday realities or evolving hybrid cosmologies. In many Oceanic cultures, for example, one finds climate change understood through a blending of Indigenous and Christian cosmologies, an amalgamation that draws together elements from both. Similarly, in the Peruvian highlands, Catholic-Andean beliefs and traditions have merged to form a blended cosmology that embraces both mountain deities and the Trinitarian God of Catholicism (Paerregaard, 2020). Subaltern voices from within Western cultures, people whose knowledge is not necessarily defined by science, also tell their stories about the everyday realities to which the idea of climate change points. For example, cultural geographers Hilary Geoghegan and Catherine Leyshon show how climate change is constructed through local memory, observation, and conversation amongst Cornish farmers on the Lizard Peninsula in Cornwall, England. It is in these "often excluded, complex and 'unquantifiable' relationships with climate and landscape", they claim, "that people make sense of and respond to climate change" (2012: 64). To understand how climate and the ways it may change relate to the everyday politics of people's lives requires listening carefully to subaltern voices wherever they may be found.

BOX 6.1 THE TRICKSTER AND CLIMATE CHANGE

Many societies nurture different stories and myths for guiding them through periods of environmental change and social disruption. These myths may appear irrational or mysterious to Western minds, yet they open up new cognitive and imaginative resources for understanding a changing and bewildering world. One such story is that of "the trickster", one of the oldest and most widespread of mythological and literary figures. The trickster figure is present in mythologies and folk traditions ranging from those of Ancient Greece and Rome, to Norse, Native American, Siberian, African, and Caribbean cultures. The trickster is typically portrayed as an "other-than-human" or "more-than-human" personality. The trickster always appears as a character that uses guile and secret knowledge to challenge authority, playing tricks on others. Although not embodying a fixed moral orientation of either vice or virtue, the trickster seeks to disrupt and transform taken-for-granted realities, "an amoral practical joker who wanders about playing pranks on unsuspecting victims. . . . With all the fluctuations, certain things about the trickster are predictable: he is always a wanderer, always hungry, and usually oversexed" (Velie, 1991: 44).

The trickster character may play different roles with respect to climate change, not least in subverting human presumptions and ambitions about knowing and controlling the climate future. The trickster introduces the virtues of humility and modesty into storytelling and suggests that human actions in the world will always exceed the ability to predict their effects. Human desire for control and mastery is tempered by the trickster. Take the trickster figure of Raven, who features widely in the North Pacific coastal cultures of America and Asia (Figure 6.1). This mythical figure has selfish, ignoble, and insatiable appetites and acts as a mirror for humanity by reflecting people's relations with the environment (Thornton & Thornton, 2015). Raven challenges the illusion of control that is promised by scientific knowledge and geoengineering technologies.

Elsewhere, from the Marshall Islands in western Oceania, Peter Rudiak-Gould (2013: 36–37) draws attention to the subaltern narrative about what he calls "modernity the trickster". He observes local responses to climate change being rooted in a historical narrative of self-inflicted cultural decline and seduction by Euro-American modernity. In Marshallese culture, modernity imported from the West is perceived as "a trickster" that has disrupted not just social relations and customs but also perturbed environmental conditions inherited from the past. There is only so much the Marshallese can do against this trickster figure intruding from outside.

Read:

Thornton, T.F. and Thornton, P.M. (2015) The mutable, the mythical and the managerial: Raven narratives and the Anthropocene. *Environment and Society: Advances in Research*. 6(1):66–86; Also Rudiak-Gould, P. (2013) *Climate Change and Tradition in a Small Island State: The Rising Tide*. Abingdon: Routledge. [especially Chapter 1].

(Continued)

Figure 6.1 Raven, a trickster figure in many North Pacific cultures and signifier of environmental change.
(*Source*: Glenn Rabena, British Columbia; reproduced from Thornton & Malhi, 2016).

Temporalities

Subaltern voices do not just unsettle assumed Western and scientific realities of what sort of phenomenon climate change is – its ontology. They may also challenge the taken-for-granted nature of time by which changes in climate are measured and predicted in Western scientific worldviews. When scientists claim to predict the temporal unfolding of climate, they appropriate the future according to a logic not necessarily shared by subalterns. Temporality in modern Western culture is conventionally understood as the chronological progression of linear time through the categories of past, present, and future. For example, Hubert Lamb used this standard tripartite division of time in his classic two-volume text, *Climate Change: Present, Past and Future* (1972, 1977). Indeed, I have used these same temporal divisions as a simple structure in this book.

Science is premised on clockwork time, benchmarked to the radiation characteristics of the caesium-133 atom.[2] Standard scientific accounts of climate change therefore partition time in this way. Precise dating techniques for fixing past climates (see Figure 2.2) and predictions of future climate for specific years such as 2050 or 2100 are the staple of climate science. This operates with the logic of "time's arrow", in which the future arrives sequentially always in fixed steps. But temporality may also be understood psychologically and sociologically with respect to people's changing *perceptions* of time and to their culture's changing social *organization* of time. For example, the Western perception of time underwent significant change between the later Middle Ages and the early modern period, and the social organisation of time in the Western world was transformed in the nineteenth century by accelerated human mobility.

Fundamental to the idea of climate change is not just time but the related question of the future (Brace & Geoghegan, 2011). The future, with its inevitability and yet attendant uncertainty, is troubling to the human imagination (see Chapter 10). It is a "cultural fact", invented and imagined in different ways in different cultures. There may be value in unsettling the simple linear flow of time assumed by the paleoclimate sciences – with respect to the past – and by climate modelling – with respect to the future. Accounts of the climatic past – or prophecies of the climatic future – need not be limited to chronological or clockwork time. Indeed, there is a tradition in Western philosophy of questioning the view that time is "a natural medium in which matter and life are framed, rather than a dynamic force [involved] in their framing" (Grosz, 1999: 3). If time is not "a natural medium" then its multiple expressions need to be engaged, not least in relation to climate change.

When subalterns engage with the idea of climate change, they may question the assumed scientific conventions of temporality. Ethnographers and anthropologists encounter multiple temporalities as they listen to and engage with subaltern understandings of what climate change signifies. Different perspectives on time and causation will prompt different questions than simply the scientific one about how did – or how will – climate change. For Indigenous scholar-activists such as Kyle Whyte the most important questions about climate change lie in the past. They are questions about responsibility for past actions and relate to violations of land, resources, and peoples by colonial settlers (Whyte, 2020). If the emphasis is on the future, then questions arise about *whose* future is being envisioned and predicted and by whom. Subalterns frequently resist what they see as the appropriation of their futures by numerical predictions emanating from climate models. For them, the climatic future is a place of contestation (Pasisi, 2019). It is not an "already-reality" that simply awaits its unveiling through the operation of predictive models operated by Western scientists (see Chapter 10).

Beyond these different ways of engaging with linear notions of time, subalterns bring other temporalities to bear on the idea of climate change. For example, understanding of climate change in Australia's Northern Territory among the Yolŋu people rests on a seasonal – or cyclical – temporal register, rather than on a linear one (O'Brien, 2016). Aboriginal notions of time are not tied to the unfolding of the Gregorian calendar. Weather seasons do not follow clock time but represent the cyclical flow of events in the natural world to which cultural life adapts (Figure 6.2). These Australian Aboriginals live with "a temporal flexibility that allows for seasons which don't happen at an exact time, but come nevertheless in cyclical sequence" (O'Brien, 2016: 39). Changes in climate are therefore experienced quite differently to how a scientist might account for them. They are experienced not as incremental increases in annual temperatures or even as interruptions to "normal" weather through extreme events. Changes in climate are experienced as adjustments or disturbances to cyclical rhythms of weather, land, labour, emotions, and rituals, which represent the integration of culture and climate, of human and non-human materialities.

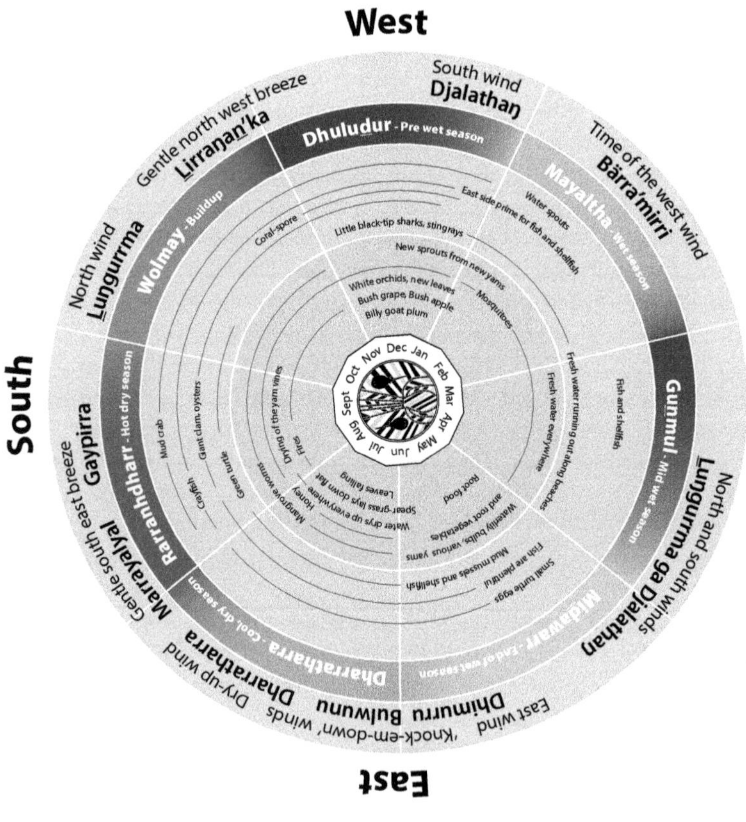

Figure 6.2 The Gurruwilyun Yolŋu Seasonal Chart, Northern Territory.
(*Source*: This image is owned by Dr Kathy Guthadjaka, AM. Reproduced with permission from the Northern Institute, Charles Darwin University, Australia).

Also deviating from Western notions of linear time are Inuit philosophies of the future. These philosophies shape how Arctic peoples imagine their relationship with climate and its changes. Inuit observations, instincts, and intuitions of change are attuned to present and immediate future conditions. It is to *these* prospective changes – not predictions years or decades into the future – that plans, decisions, and adjustments are made (Bates, 2007). This is not to say that Inuit deny there are longer-term environmental changes afoot. It is rather that they engage with changes in physical conditions in ways that are more immediate and provisional. These are modes of living that over many generations have proven themselves adaptive in a volatile and unpredictable environment. For Inuit, future uncertainties are not to be eliminated through striving for barely attainable knowledge, least of all knowledge offered from numerical models which – in their view – fail to represent local realities.

Resistances

One of the earliest expressions of political resistance to the dominant scientific framing of climate change came from two Indian social scientists, Anil Agarwal and Sunita Narain. Working in the Centre for Science and Environment in Delhi they wrote a powerful pamphlet in 1991 in which they drew the distinction between what they called "luxury emissions" and "survival emissions" (Agarwal & Narain, 1991). A scientific approach to understanding climate change regards all molecules of carbon dioxide equally. But as Agarwal and Narain pointed out, while the physical properties and climatic effects of two molecules of carbon dioxide are identical, their moral weight can be radically different. Ethically, it matters whether a molecule of carbon dioxide is emitted by an Indian peasant farmer or a Russian tourist flying to Black Sea resorts. Science is blind to the difference between luxury and survival emissions. Subalterns – and others who frame climate change first as a question of justice – are not (Luke, 2018).

In many senses, Agwaral and Narain laid the path for what later emerged as the climate justice movement. Many subalterns' engagements with the idea of climate change – indeed their very understanding of what climate change is – resists the dominant framings, discourses, and representations of climate change that emanate from centres of power, whether these be global, national, or networked. To understand these multiple sites and motives of political resistance a number of different dimensions of justice come into play: Representational, epistemic, intersectional, and restorative. We will look briefly at each of these.

The representational politics of climate change was alluded to earlier, using the example of the *Sydney Morning Herald's* representation of climate change and the islanders of Tuvalu. Such condescending forms of representation – repeated in other regional settings around the world – often provoke acts of resistance by subalterns. For example, in her study of climate change narratives emerging from Oceania, Hannah Fair shows how new political and activist subjectivities are created by climate resistance movements such as the Pacific Climate Warriors (Fair, 2018). In this particular example – and at other sites of resistance to dominant Western framings of climate change – religious beliefs inform local understandings of climate change agency and political action (see Box 8.1). In similar vein, Bronwyn Hayward and her colleagues from Oceanic small island nations resist narratives of climate despair and associated claims that it is "too late" to confront climate change (Hayward et al., 2020). Rather, they focus attention on the many efforts already initiated among island peoples – particularly those led by women and youth – that are "informed by distinctive community values of *Vai Nui* or *Fonofale* ('interconnected well-living')". These values have sustained island societies "through traumatic histories of colonization, racism, and violence, and are still positioned to support communities suffering now, and when facing future risks" (ibid, p. 1).

The dimension of *epistemic justice* has been alluded to previously when thinking about the dangers of extractivism within integrationist perspectives on subaltern knowledges (see Box 6.2). Giving due recognition to experiential and other non-scientific descriptions of climate change is important if communities and individuals are to be allowed "to work within their own culturally specific socio-natural entanglements to produce [political and social] change" (Rice et al., 2015: 260). Foregrounding subaltern voices on climate change can also be understood as a form of knowledge decolonisation. It empowers diverse communities of action to privilege relational, moral, and ethical considerations over techno-scientific ones.

BOX 6.2 NICOLE KLENK (B.1977) – EPISTEMIC JUSTICE

Nicole Klenk is a political ecologist who studies the politics of climate knowledges. She is currently Associate Professor at the Department of Physical and Environmental Sciences at the University of Toronto, Scarborough, Canada. Her academic background is in botany and forest ecology. After gaining a BSc and MSc from McGill University, her PhD was awarded in 2008 from the University of British Columbia for a study of "The Ethics and Values Underlying the 'Emulation of Natural Disturbance' Forest Management Approach in Canada: An Interdisciplinary and Interpretive Study". Her PhD thesis showed how ethics shapes the interpretations of forests and she advanced a pragmatic approach for adopting a more holistic approach to knowledge-making in forestry science.

During her graduate research, Klenk became interested in the role of science in addressing complex environmental problems such as climate change. Her work subsequently expanded to investigate the role of (environmental) science in society and the ethics, politics, and governance of knowledge creation and use. Klenk's work sits in the tradition of critical geography and political ecology, drawing inspiration from science studies, post-structuralist political theory, and early American pragmatism. She has applied her thinking to the areas of forestry, biodiversity conservation, and especially climate change adaptation, working in settings across the Americas, from Colombia to the Canadian Arctic.

Klenk exposes how and why different climate knowledges get authorised in different contexts, who gets to control such knowledges, and how this political dynamic changes over time. In a series of important papers, Klenk has argued against the "knowledge integration imperative", which is present in much transdisciplinary environmental science. This imperative is the idea that Indigenous and subaltern knowledges can and should be integrated with Western climate science. For Klenk, such a move obscures the politics of scientific knowledge – including the friction, agonism, and power relations inherent in knowledge coproduction. Subaltern knowledges need understanding on their own terms. Her work makes an important contribution to the decolonisation of climate knowledge by resisting the appropriation of subaltern knowledges by Western natural and social sciences.

Read:

Klenk, N. and Meehan, K. (2015) Climate change and transdisciplinary science: Problematizing the integration imperative. *Environmental Science & Policy*. 54: 160–167; Latulippe, N. and Klenk, N. (2019) Making room and moving over: Knowledge co-production, Indigenous knowledge sovereignty and the politics of global environmental change decision making. *Current Opinion in Environmental Sustainability*. 42: 7–14.

Subaltern voices also bring an added dimension to social movements concerned with *intersectional justice* and climate change. We looked briefly at this in Chapter 5. New social movements, seeking transformative change in response to climate change, increasingly build their case around the connections between social, economic, gender, and racial justice (Hathaway, 2020). Subalterns add a postcolonial perspective to such justice-based arguments. For example, Kyle Whyte has been a powerful voice arguing for Indigenous climate justice for First Nation Americans. For Whyte and other Indigenous activists, climate change offers a discursive site of resistance that can be used to promote Indigenous collective self-determination and to advance the case for restorative justice (see later).

Scholars such as Emilie Cameron, Michael Bravo, and Sheila Watt-Cloutier make similar arguments with respect to Inuit-inhabited Arctic territories. Imported narratives of climate change in the Arctic – whether borne by Western scientists or foreign NGOs – are frequently expressed in terms of disappearing ice and local crises in ecosystem resilience. Such narratives narrow the questions that can be asked about what is at stake with climate change. They offer only a limited space within which subaltern citizenship can or should be expressed. Rather than uncritically adopting such narratives, Inuit climate campaigners such as Sheila Watt-Cloutier (see Box 6.3) re-express what lies at the heart of climate change, namely the legacies of colonialism, political exclusion, economic marginalisation, and cultural violation (Bravo, 2009; Cameron, 2012).

BOX 6.3 SHEILA WATT-CLOUTIER (B.1953) – INUIT CLIMATE ACTIVIST

Sheila Watt-Cloutier was born in Kuujjuaq, in the Nunavik region of Northern Quebec, Canada, in 1953 and raised there traditionally before she was sent away for school in Nova Scotia and Churchill, Manitoba. Returning years later to Nunavik, she worked as an Inuktitut health translator and as a counsellor in the education system, in both roles seeking to improve educational and health conditions among Inuit. For nearly 20 years Watt-Cloutier was a political representative for Inuit people at regional, national, and international levels. She became Corporate Secretary of a Canadian Inuit land-claim organization and in 1995 was elected President of the Inuit Circumpolar Council (ICC) Canada and re-elected in 1998. In this position, she served as the spokesperson for Arctic Indigenous peoples in the negotiation of the

Stockholm Convention banning the manufacture and use of persistent organic pollutants. In 2002, she was elected International Chair of ICC, a position she held for four years.

Watt-Cloutier became increasingly engaged in campaigning for the rights of Inuit people. In 2005, along with 62 Inuit Hunters and Elders, she launched one of the world's first international legal actions on climate change. This was a petition to the Inter-American Commission on Human Rights (IACHR), alleging that unchecked emissions of greenhouse gases from the United States had violated guaranteed Inuit cultural and environmental human rights. Although the IACHR decided against hearing her petition, in 2007 the Commission invited Watt-Cloutier to testify with her international legal team at their first ever hearing on climate change and human rights.

Watt-Cloutier has received numerous awards and honours for her work, including as an advisor to Canada's Ecofiscal Commission developing new environmental policies. She continues to speak and advocate on behalf of her people and their political and environmental rights, especially in the context of climate change. Although she gained recognition and influence through her climate change campaigning, in her 2015 autobiography she explains, "I came to this particular mission through the concern I had for our people and my great desire to protect the Inuit way of life" (Watt-Cloutier, 2015: 316).

Read:

Watt-Cloutier, S. (2015) *The Right to be Cold: One Woman's Fight to Protect the Arctic and Save the Planet from Climate Change*. Minnesota: University of Minnesota Press. See also her speeches and interviews on receipt of the 2015 Right Livelihood Award: www.rightlivelihoodaward.org/laureates/sheila-watt-cloutier/ [Accessed 20 July 2020].

These subaltern concerns form core elements of what climate change means for many people but are elements that cannot be captured by scientific measurements or represented in model predictions. They fuel a fourth axis of resistance, namely the demand for *restorative justice*, sometimes referred to as reparative or corrective justice. Modernist and transformative discourses of climate change tend to focus on the distributive and/or procedural dimensions of justice. Who suffers? And who gets to decide? But the subaltern's call for restorative justice is seeking redress for vulnerabilities to climate change that are a result of previous injustices. A focus on restorative justice is to ensure that it is not just the victims of climate change who are remembered. It is also to make sure that those who perpetrated past oppressions are not ignored. Restorative justice demands that the interests and perspectives of postcolonial or subaltern communities in the Global South – the primary victims of climate injustice in the eyes of many – are heard and are made to count (Bhavnani et al., 2019). Restorative justice also seeks to protect the needs and rights of future generations and, in some accounts, those of non-human entities as well. The claims for restorative justice resonate especially powerfully for those whose ontologies are relational.

These four manifestations of subaltern resistance to the techno-economic-scientific narration of climate change have resulted in real political effects. For example, the creation in 2013 at COP-19 of the Warsaw International Mechanism (WIM) for Loss and Damage was shaped by the views of many subalterns from developing countries who feared climate change impacts may exceed their ability to adapt. Although calls for "climate reparations" were unsuccessful, the WIM requires richer nations to assist those suffering "loss and damage" from climate change by promoting knowledge transfer and enhanced capacity-building. And two years later, the views of subalterns from Small Island Developing States were undoubtedly significant in securing the insertion of the 1.5°C temperature threshold as an aspirational target in the 2015 Paris Agreement. In these cases, subaltern voices are appropriating scientific language and framings in order to advance their political claims, which are grounded in "more-than-science" reasoning. One might also point to the growing use of litigation through the courts as a manifestation of subaltern power, plaintiffs often being young people, ordinary citizens, or the claimed victims of climate harms (see Chapter 9). Although successful climate litigation remains relatively rare, the rising number of court cases points to an emerging locus of climate governance where subalterns might find it possible to exercise power otherwise denied them.

Chapter summary

This chapter makes the case for understanding the idea of climate change from the perspective of subalterns, through listening to the experiences, voices, and stories of peoples and cultures who may have little or no access to the levers of power or to elite discourse. Rather than starting with scientific accounts of climate change, listening to subaltern voices starts with the recognition of ontological diversity and political subjectivity. As the idea of climate change travels around the world, it is reimagined iteratively according to different situated knowledges and subjective accounts of meaning. When encountering different ontologies, cultural identities, and historical trajectories, the standard scientific account of climate change may be resisted or else translated into locally meaningful narratives. The argument of this chapter is that the meaning and relevance of climate change for many people around the world can only be understood from this geographically sensitive stance. Merely communicating or downscaling the science of global climate change is not enough.

Rather than encounters between different climate knowledges leading to a flattening or an integration of knowledge, the opposite happens. The universal idea of climate change becomes fragmented, refracted, and subverted. Subaltern interpretations of climate change then empower local forms of political mobilisation and resistance. To understand the idea of climate change "from below" requires one to be a better geographer or a better anthropologist, not a better scientist. Answering questions about how climate change relates to the politics of people's everyday lives does not require better climate models with more precise predictions. It requires careful listening to the realities and concerns of subaltern voices.

Notes

1 This vignette of Mr Leboi's testimony is adapted from Sara de Wit's PhD thesis, *Love in Times of Climate Change: How an idea of Adaptation to Climate Change travels to northern Tanzania*, awarded by the University of Cologne in 2016.

2 For the atomic clock, agreed in 1967, one astronomical second is defined as 9,192,631,770 periods of the radiation corresponding to the transition between the two hyperfine levels of the ground state of the caesium 133 atom.

Further Reading

A good starting point for understanding how anthropologists foreground subaltern perspectives on climate change is **Peter Rudiak-Gould's** *Climate Change and Tradition in a Small Island State: The Rising Tide* (Routledge, 2013). In similar vein is **Tim Leduc's** *Climate, Culture, Change: Inuit and Western Dialogues with a Warming North* (University of Ottawa Press, 2010), which highlights the challenges posed by Inuit knowledge to Western climate research and Canadian climate politics. A classic in this genre is also still very much worth reading, namely **Julie Cruikshank's** *Do Glaciers Listen? Local Knowledge, Colonial Encounters and Social Imagination* (University of British Columbia Press, 2005). An excellent example of a narrative account of what climate change means to marginalised rural communities in Global North societies is **Bradon Leap's** *Gone Goose: The Remaking of an American Town in the Age of Climate Change* (Temple University Press, 2018). And for a comparable account for rural dwellers in coastal Bangladesh read **Manoj Roy, Joseph Hanlon**, and **David Hulme's** *Bangladesh Confronts Climate Change: Keeping our Heads Above Water* (Anthem Press, 2016). In similar vein, *Kivalina: A Climate Change Story* (Haymarket Books, 2011) by **Christine Shearer** is a short but powerful account of the battle faced by the Inupiat residents of Kivalina, Alaska to relocate their village threatened by a rising ocean. *The Social Life of Climate Change Models:*

Anticipating Nature (Routledge, 2013) is a collection of essays edited by **Kirsten Hastrup** and **Martin Skrydstrup** that helpfully juxtapose different ways of knowledge-making about climate – both scientific and subaltern. The studies emphasise how deeply social are all such efforts, whether emanating from coastal India, the Cook Islands, mountainous Tibet, or the High Arctic. For a personal account of the struggle to make subaltern voices on climate change heard on the global stage **Sheila Watt-Cloutier's** *The Right to be Cold: One Woman's Fight to Protect the Arctic and Save the Planet From Climate Change* (University of Minnesota Press, 2015/2018) is to be recommended. Finally, **Tariq Jazeel** and **Stephen Legg** offer a collection of edited essays in *Subaltern Geographies* (University of Georgia Press, 2019) that offer valuable insights for thinking more broadly about subaltern perspectives on climate change.

QUESTIONS FOR CLASS DISCUSSION OR ASSESSMENT

Q6.1: Can scientific accounts of climate change be reconciled with – or integrated into – subaltern understandings of the phenomenon that derive from radically different ontologies and political histories?

Q6.2: In what ways do subalterns use Western science to advance their claims for climate justice? Is this appropriation of science inevitable and/or desirable?

Q6.3: "If we understand climate change through various kincentric perspectives, then a relational tipping point was probably crossed years ago through the operations and impacts of colonialism, industrialization, and capitalism" (Whyte, 2020: 5). If this is so, is it too late to stop climate change?

Q6.4: What attention should be given to Indigenous knowledges of climate within the IPCC's global assessments of climate change? If this goal is pursued by the IPCC is there a danger of epistemic appropriation of Indigenous knowledge by science? (See Box 6.2.)

Q6.5: In what ways – if any – is the idea of climate change gendered?

7

ARTISTIC CREATIVITIES

. . . Climate change reimagined . . .

Introduction

Just over 200 years ago unusual weather patterns were experienced across the northern hemisphere over a number of years. The cause of the climatic disruption was mysterious. No one at the time – or indeed for well over 150 years after – made the connection between drought in China; warmth in Japan; a series of cold, wet summers; harvest failures in Europe and North America; and a massive volcanic eruption to the east of Java. The eruption in April 1815 of Mount Tambora on the island of Sumbawa killed between 75,000 and 100,000 people in the surrounding islands, either immediately due to the explosion and debris or else due to starvation and disease in the months following. As the ash cloud spread around the world, Europe's summer in 1816 was cold, wet, and gloomy, prompting the epithet "the year without summer".

It was during this dark summer of 1816 that the young English author Mary Shelley wrote her famous novel *Frankenstein: or, The Modern Prometheus*, published for the first time in 1818 (Shelley, 1818). Her gothic fiction was a response to a dare from Lord Byron to his literary colleagues – holed-up on the shores of Lake Geneva during the miserable weather – to write a ghost story. *Frankenstein* is regarded by many as the first work of science fiction and it continues today to influence attitudes to emerging science and technology (Guston et al., 2017). In *British Romanticism, Climate Change and the Anthropocene* (2017), David Higgins uses the texts of Byron and the Shelleys – Mary and her husband Percy Bysshe – from the period 1816–1819 to reflect on the impact of this climatic disturbance, temporary as it was, on **Romantic** literary culture. Higgins shows how writers at the time responded to climatic disturbance with foreboding but creative energy.

Inspired by Byron's literary dare in the "summer-less summer" of 1816, curator and aestheticist Dehlia Hannah recently embarked on a creative project of her own. Romantic authors such as Bryon and the Shelleys offered a creative response to the year without summer. Two hundred years later Hannah asked what might a "year without a winter" look and feel like in aesthetic terms? She wanted to explore how the relentless cognitive and imaginative challenges of the idea of climate change in today's world might find expression in multiple cultural registers. How might artistic and literary engagements with *today's* destabilizing climatic patterns open up new planetary imaginaries? How might they offer new perspectives on humanity's relationship with its environment?

Hannah's *A Year Without a Winter* (2018) brings together science fiction, history, visual art, and exploration. The 1815 eruption of Mount Tambora had enveloped the northern hemisphere sub-tropics in a cloud of ash that temporarily reduced global temperature by about half a degree Celsius. Two hundred years later in 2016 – the (then) hottest year globally on historical record – Hannah assembled four renowned science fiction authors in the experimental utopian town of Arcosanti in the Arizona desert. She asked them each to respond in prose to this new climatic disturbance, a latter-day Byronic literary dare. In *A Year Without a Winter*, Hannah presents their four stories, alongside other critical essays, extracts from Shelley's masterpiece, exhibitions, postcards, and dispatches from expeditions to extreme geographies. Hannah also created her own seasonal cosmographic, which depicted each of the 12 calendar months in iconic form, connected through imaginary geometric gridlines (Figure 7.1). Her project of *A Year Without a Winter* was an experimental aesthetic appreciation of the idea of climate change. It was not an exercise in science communication but a reimagining of what the idea signifies to creative minds.

***

The last quarter century has seen a growing engagement by the creative arts with the idea of climate change (Galafassi et al., 2018). This has been part of a much broader rapprochement in Western culture between the sciences and the arts – a dissolving of the "two cultures" framing of Western intellectual heritage proposed by C.P. Snow in his eponymous 1959 lecture. A growing artistic engagement with climate change also parallels the "creative turn" in geography (see Box 7.1 and the work of Harriet Hawkins) and the emergence in the 1990s of the sub-discipline of environmental humanities. Part of this creative move in relation to climate change can be explained by a belief in art's social agency, defined by David Maggs and John Robinson in *Sustainability in an Imaginary World* (2020) as the capacity of art to advance social goals, increase knowledge, stimulate empathy and care, and change behaviour. This search for art's instrumental social agency has been particularly alluring for the case of climate change. Publics have, it is claimed, been obstinate in failing to respond "appropriately" to scientific messages of changing climatic realities. Creative entrepreneurship has therefore blossomed, seeking to communicate climate science more "effectively" by aestheticizing it (Sommer & Klöckner, 2021). In such initiatives, art is appropriated as a handmaiden to science.

Figure 7.1 The cosmographic frontispiece for Dehlia Hannah's *A Year Without a Winter*. Each icon stands for a calendar month, reading anti-clockwise from January (top). (*Source*: Hannah, 2018; reproduced with the author's permission).

BOX 7.1 HARRIET HAWKINS (B.1980) – CREATIVE GEOGRAPHIES OF CLIMATE

Harriet Hawkins is Professor of GeoHumanities in the Department of Geography at Royal Holloway University of London and Co-Director of the Centre for the Geo-Humanities. She gained a BA in geography and an MA in landscape and culture from the University of Nottingham, which also awarded her PhD in 2006 for a thesis titled "Geographies of Art and Rubbish: An Approach to the Work

(Continued)

of Richard Wentworth, Tomoko Takahashi and Michael Landy". She is currently editor of the journal *Cultural Geographies* and a recipient of a major five-year European research grant to develop novel and creative approaches to engaging subterranean worlds, for example imagining the future of underground cities, exploring the complexities of extraction, and conserving the heritage of underground landscapes around the world.

Hawkins' research is focused on the advancement of the geohumanities, a research field that sits at the intersection of geographical scholarship with arts and humanities scholarship and practice. *Empirically* she explores the geographies of art works and art worlds. *Theoretically* she is interested in the elaboration from a geographical perspective of core humanities concepts of aesthetics, creativity, and the imagination. Hawkins' research includes collaborating with artists and institutions to create artwork and to curate exhibitions and events. In addition to her own written research, she has produced artists' books, participatory art projects, and exhibitions with individual artists and a range of international arts organizations, including Tate, Arts Catalyst, Iniva, Furtherfield, and Swiss Artists in Labs.

Hawkins has applied her empirical and theoretical work to the idea of climate change and the different modes of creative practice that engage with it. In particular, she has examined how audio and visual art addresses the enduring problems of climate change communication. She has argued that lessons learnt about sensory experience, affect, and emotions might be more widely applied to the analysis of cultural forms – from literature to films – and their role in climate change communication. Hawkins' work sits within the evolving field of participatory and interactive art that enrols and empowers citizens. These art forms often feature community and site-specific practices that take up local climate and environmental issues, whether through story-telling or through gathering and curating living archives and artefacts of change.

Read:

Hawkins, H. (2013) *For Creative Geographies: Geography, Visual Arts and the Making of Worlds*. Abingdon: Routledge; Hawkins, H. and Kanngieser, A. (2017) Artful climate change communication: Overcoming abstractions, insensibilities, and distances. *WIREs Climate Change*. 8(5): e472; Hawkins, H. (2019) Geography's creative (re)turn: Toward a critical framework. *Progress in Human Geography*. 43(6): 963–984.

This chapter puts forward a different perspective. The goal of artistic engagement with climate change is not to inspire or support art that carries "the message". After all, the idea of climate change *has* no single timeless meaning waiting to be revealed by an artist. Climate and its changes depicted by science has no mimetic representation, whether by computer simulation, photography, performance, creative fiction, or film. As Garrard and colleagues argue in the case of literature, "ecocriticism should not become the literary department of the IPCC" (Garrard et al., 2019: 207). No. For those seeking to preserve the integrity of arts practices, creative engagement with the idea of climate change is about using different cultural media to disrupt taken-for-granted representations of climate change. It is about "thickening" people's understanding of climate change rather than disciplining it.

A question frequently asked of contemporary climate art is whether it has – or whether it *should* have – the power to change people's beliefs, attitudes, and behaviours that scientific representations of climate change seem not to possess. But the question is badly posed. For Maggs and Robinson this is the wrong way – or at least it is not the only way – to think about the relationship between art and climate change. For them, art's social agency resides more in its ability to evoke a more-than-rational, transformative experience. Art has a "capacity to foster a sense of otherness, a sense that the categories and conceptions by which we live are just that, categories and conceptions, limited versions of limitless possibilities" (Maggs & Robinson, 2020: 46). Art that is inspired by the idea of climate change should therefore be just as likely to *problematise* scientific realities, attitudinal changes, or prospective solutions as it is to successfully implant pre-packaged concern or motivations for predefined behavioural change. Artistic processes should move climate discourse away from scientific imperatives and instead to "reauthor the world" by exploring existential meanings, identities, and purposes beyond delivering net-zero emissions (Maggs & Robinson, 2020). Emanating from multiple forms of artistic engagements should be invitations to imagine climate change through different cultural registers and to be moved in ways that science is unable to provoke (Figure 7.2).

The chapter draws on the disciplines of *creative geographies, environmental humanities,* and *museum studies* to reflect on the multiple ways in which creative artists have responded to the provocation of climate change. It is organised in three main subsections. The first two deal with artistic engagement with climate change through, respectively, textual and visual media, while a third subsection considers the role of museums and their potential to give creative form to the idea of climate change for diverse public audiences. Climate art engages the full range of human faculties to provoke reflection on profound questions prompted by the idea of climate change: The interdependence between climate and culture; how to live with uncertainty; the future to be aspired to;

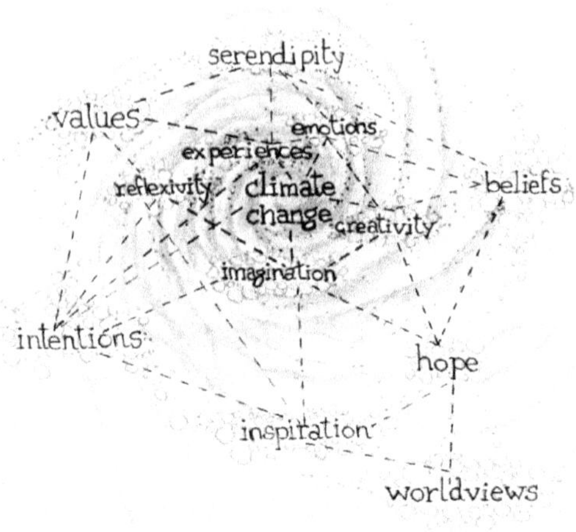

Figure 7.2 An artistic illustration of climate change operating as a creative cultural force prompting inquiries into a wider range of human subjectivities.
(*Source*: Diego Galafassi).

the life to be admired; the responsibilities people have to others, both human and non-human. Artists engage with the idea of climate change in order to explore the relationship between climate, culture, and human agency, without being subordinated to a scientifically prescribed reality. It is for this reason that I position this chapter on artistic creativities as a "more-than-science" approach for understanding the idea of climate change.

Textual

Given the salience of climate change in public life over the past quarter century, the proliferation of literary texts engaging the idea of climate change is unsurprising. Climate change touches all aspects of human cultural life – material, performative, and imaginative. It is inescapably a literary phenomenon as much as it is an object of scientific inquiry. How climate change is written into literary culture using the imagination matters as much as what science may disclose to us about the workings of the physical climate. The idea of climate change has permeated the creative writing of many new and established authors around the world – whether directly or indirectly – and whether in novels of various genres, short stories, poems, plays, or film scripts. This subsection examines some of these modes of writing climate change.

The rise of "the climate change novel" has been a remarkable feature of the twenty-first century (Goodbody & Johns-Putra, 2019). It has provoked the emergence of what some have referred to as a new genre of writing: "Climate fiction" or Cli-Fi (see Box 7.2 for further elaboration). Cli-Fi sits within the longer literary tradition of ecocriticism, which in turn has its origins in Romantic poetry and nature-writing. But while ecocriticism might have originally developed as a way to connect the reader to a sublime nature through a critical rendering of environmental writing, since the 1990s it has adopted a more sceptical and questioning tone. An ecocritical approach to reading "climate literature" is not one that necessarily cheerleads for climate science. Nor is it one that suggests ever more persuasive plots, tropes, and styles in order to convince a reluctant public to alter their behaviour because of the threat of climate change. (Although for a different perspective on the purpose of ecocriticism, see Schneider-Mayerson et al., 2020.) Through a close reading of texts, ecocriticism is more likely to expose ambivalence about what climate change *is*, people's anxieties about the future, and the nature of the predicament climate change presents us with.

BOX 7.2 CLIMATE FICTION . . . OR "CLI-FI"

The abbreviation "Cli-Fi" was a 2007 neologism of the Taiwan-based journalist Dan Bloom. By analogy with Sci-Fi, Cli-Fi suggests a new genre of literary and film fiction that has emerged in response to the idea of climate change, a genre that has certainly gained in cultural salience over the past decade. In 2016 Michael Svoboda surveyed a corpus of more than 60 Cli-Fi films, including major theatrical releases, smaller festival films, and made-for-TV movies. And, more recently, literary scholars Axel Goodbody and Adeline Johns-Putra (2018) have compiled a "Companion" to Cli-Fi literature comprising a series of commissioned essays commenting on 24 selected Cli-Fi novels and five Cli-Fi films.

Goodbody and Johns-Putra define Cli-Fi as a distinct body of cultural work that explores the idea of climate change "not just in terms of setting, but with regard to psychological and social issues, combining fictional plots with meteorological facts, speculation on the future, and reflection on the human-nature relationship" (p. 1). This leads them to select a series of exemplar texts grouped into five subgenres, namely

(Continued)

realist narratives set in the present or near future, *speculative future fiction* (dystopian and apocalyptic narratives), *genre fiction* (thriller, crime, conspiracy, social satire), *children's film and young adult novels*, and *literary modernism*. Their "Companion" also includes reflections on what they term "proto-climate-change fiction", exemplified by the novels of J.G. Ballard (*The Drowned World*, 1962), Max Frisch (*Man in the Holocene*, 1980), and Ignatio Brandão (*And Still the Earth*, 1985). These three works pre-date public concerns about climate change triggered by scientific explanations of human-caused global heating. They suggest that writers have always approached the idea of climate change in the light of questions about human agency and responsibility.

Svoboda's examination (2016) of filmic Cli-Fi offered a different typology for films' engagement with the idea of climate change. His survey showed that two tropes dominated the narrative plots of these films: The visual drama of extreme weather events – as in the *Sharknado* (2013–2015) film series and *Noah* (2014) – and the possibility of a new ice age on Earth, as for example in *The Day After Tomorrow* (2004) and *Snowpiercer* (2013). Svoboda suggests that in popular film the "disaster movie" is very much the go-to form for putting the idea of climate change on the big screen. He also suggests that the melodrama offered by this trope easily allows commercial Cli-Fi films to slip into parody. Almost entirely absent from his corpus were films that portrayed successful efforts to mitigate the causes of climate change and only a very few addressed ways to adapt successfully to its consequences.

Read:

Goodbody, A. and Johns-Putra, A. (eds.) (2018) *Cli-Fi: A Companion*. Oxford: Peter Lang; Trexler, A. (2015) *Anthropocene Fictions: The Novel in a Time of Climate Change*. Charlottesville: University of Virginia Press; Svoboda, M. (2016) Cli-fi on the screen(s): Patterns in the representations of climate change in fictional films. *WIREs Climate Change*. 7(1): 43–64.

Literature has a unique ability to capture the complexity and ambiguity of everyday human experience. Through the medium of story-telling, novels are able to negotiate between competing cultural values and convey contrary experiences of class, race, and gender. They are able to provoke reflection about our actions in the world in relation to the subjectivities of others and the imagined possibilities of an unknown future. Literature therefore plays an important public role in illuminating the idea of climate change. Goodbody and Johns-Putra conclude their survey of the climate change novel by observing that "Climate fiction helps to define our perception of climate change, while drawing out its social and political, philosophical and ethical implications" (2019: 245). And it does this in a way entirely different to that offered by scientific analysis or rationality. In what follows I comment briefly on the literary engagement with climate change through four genres of fiction: Realism, magical realism, speculative fiction, and **Romanticism**.

Climate change presents literary realism with a formidable challenge. The challenge is how to embrace the vast and disorienting scales of time and space over which climate change unfolds and, at the same time, how to make these scales tangible through narratives that relate to readers' everyday lives. The difficulty of the idea of climate change for classic realist fiction has been pointed out by the Indian novelist and literary critic Amitav Ghosh in his widely read book *The Great Derangement: Climate Change and the Unthinkable* (2016). Ghosh identifies what lies at the heart of the challenge of writing creatively about climate change. The success of the modern novel in its realist genre lies in capturing the everyday realities and experiences of the human subject. But the emphasis on order, uniformity, predictability, and human agency that has lain at the heart of the realist novel from the nineteenth century onwards sits uneasily with the extra-human and unsettling scalar characteristics of climate change.

These are the characteristics that prompted ecocritic Timothy Morton in 2013 to refer to climate change as a "hyperobject", by which he means something that is massively distributed in space and time relative to humans' sense of scale (Morton, 2013). How does one write a realist novel about a hyperobject that is so "un-real"? How does climate change relate to everyday human experience? This difficulty explains why much climate literature has adopted the genre of speculative fiction (see later), which allows more degrees of freedom to engage the imagination.

Nevertheless, some well-known authors have enrolled in the project of producing realist Cli-Fi, for example English novelist Ian McEwan in his satirical allegory *Solar* (2010), German author Ilija Trojanow in her novel *Melting Ice* (2011), and American writer Barbara Kingsolver's *Flight Behavior* (2012). None of these three well-publicised novels entirely succeed as great realist fiction. *Solar* engages boldly with the cultural politics of climate change, but many critics found the novel and McEwan's characterisation of its chief protagonist – the scientist Michael Beard – disappointing. Critics of Trojanow's *Melting Ice* drew attention to the aesthetic challenges of writing about climate change, pointing out the tension between the author's confessional and didactic impulses. Kingsolver's *Flight Behavior* was better received by critics and captures very well the social realities of small-town rural America and its climate scepticism. On the other hand, her frequent insertions of instructive text about climate science is laboured and places the reader too often under instruction in the classroom. In different ways, all three novels confirm Ghosh's observation about the challenge of the idea of climate change for realist fiction.

Magical realism is one subgenre of realism that has more recently attracted the attention of some literary critics with regard to climate change. Within magical realism the fictional world is still recognisably "real" but has an undercurrent of magical or fantastical elements that are considered normal. This subgenre has been explored by Ben Holgate in *Climate and Crises: Magical Realism as Environmental Discourse* (2019) where he suggests this mode of writing promises to reimagine climate change beyond the limitations of conventional realist literature. Holgate argues that magical realism is frequently grounded in postcolonial perspectives, creating fictional worlds that resist dominant Western ontologies such as scientific materialism. It also offers a biocentric focus on the interconnectedness of all existing beings. These characteristics resonate powerfully with a "more-than-scientific" approach to understanding climate change. Magical realist climate fiction therefore offers a literary strategy for reimagining climate change through subaltern minds (see Chapter 6), as indeed many Indigenous and Asian authors have pursued. Examples of magical realist climate novels according to Holgate include the award-winning *The Bone People* (1984) by Māori author Keri Hulme and Taiwan's Wu Ming-yi with his *The Man With Compound Eyes* (2013).

The challenges of writing climate change into realist fiction have meant that many authors have chosen to explore the idea of climate change through speculative fiction, which includes the genre of science fiction. These novels are situated in climate-changed futures that, most frequently, are apocalyptic, post-apocalyptic, or dystopian. (Cli-Fi very rarely offers a utopian future; see Chapter 10). Cormac McCarthy's *The Road* (2006) – although nowhere mentioning climate change – is regarded as archetypical of the genre. Ecocritics such as Greg Garrard and Ursula Heise have identified **declensionism** as the dominant trope for speculative climate fiction. In this respect fictional climate dystopias fit a well-developed pattern in environmental writing (Buell, 1995). In "decline stories", writes Heise, "the awareness of nature's beauty and value is intimately linked to a foreboding sense of its looming destruction" (Heise, 2016: 7). Apocalyptic climate dystopias offer a pathway and a timetable towards endings and extinctions. In this sense they mirror some of the more extreme public discourses about climate change and the rhetoric of declensionist social movements such as Extinction Rebellion (see Box 5.3).

Whatever else is happening in such speculative climate fiction it most certainly cannot be said to be ventriloquizing climate science. For example, one of the more widely selling Cli-Fi authors – Kim Stanley Robinson – sets his most recent novel *New York 2140* (2017) in an urban Manhattan flooded by an implausible 16-metre rise in seawater level. Although New York is mostly underwater, people still live in the upper floors of the city's buildings or else in skyscrapers newly constructed for the rich in Uptown Manhattan. The idea of climate change may easily lend itself to such extremities of the human imagination. It pushes us beyond everyday realism into unfamiliar and therefore frightening worlds. But it should also be noted that such apocalyptic fiction has psychological distancing effects on human subjectivities. It runs the risk of sapping the mobilizing energy of social movements or of slipping into cynicism and apathy.

A brief comment is also in order with regard to the relationship between Romanticism and climate fiction. It was with the Romantics, Byron and Shelley, that I started this chapter. They applied their creative geniuses to giving narrative – and poetic (see Box 7.3) – form to their perceptions of the climatic disturbances that invaded their early nineteenth-century world. David Higgins has traced this genre of climate writing back into the latter decades of the eighteenth century. Romantic writers were prescient in the way they approached the idea of climate change. They grasped the deep interpenetration and codependence of the natural and cultural worlds in ways that other authors didn't. For example, in *Queen Mab* (1813) and *Prometheus Unbound* (1820) Percy Bysshe Shelley imagined the conjoining of political and climatic bliss. As Higgins (2019: 129) observes, for Shelley "a climatic shift into an eternal spring or summer signifies the world's socio-political liberation and the perfectibility of humanity". Romantic writers offer a view of climate change that speaks to the sublime catastrophe of an unravelling climate but also to the tortuous entanglements of human agency with the climate. They therefore presage the present predicament of climate change.

Whichever genre an author adopts – and however their writing is received critically – literary representations of climate change seek to accomplish what all good writers of fiction aim for. That is, to draw readers into new imaginative worlds in which they learn more about themselves in relation to others and about their emotional, intellectual, philosophical, and spiritual capacities. Whether branded Cli-Fi or not, climate fiction is about opening up possibilities for readers to feel and imagine climate change in new ways. Whatever else it accomplishes, "the diverse storylines of climate fiction make it impossible to think about the future in a singular way" (Milkoreit, 2016: 179).

Drama

A different mode of textual engagement with climate change is the play. Stage drama is an apt medium for dialogue and debate about unwelcome – but recognisably human – predicaments. Even though theatres offer controlled and artificial environments designed for the presentation of human behaviours, they nevertheless offer a suitable vehicle for exploring the complex predicament of climate change. The stage offers a powerful means for representing challenging personal and emotional entanglements with the idea of climate change and the fears and aspirations these entanglements provoke. Playwrights of the twenty-first century have grappled with how to tell the story of climate change through stage drama and through dance and music. The idea of climate change has now been "performed" on stage or in public spaces in many countries, with examples ranging from the Fijian dance-drama-musical *Moana* – literally "the rising of the sea" – to *3rd Ring Out* by the British director Zoë Svendsen and to participatory drama in rural communities in Kenya.

Theatre scholar Stephen Bottoms analysed four such stage plays that appeared on the London stage around the year 2010: *The Contingency Plan*, *The Heretic*, *Greenland*, and *Earthquakes in London*. Building on a long British tradition of politically engaged theatre, these plays dramatized the often-difficult relationships between scientists, politicians, activists, and lay publics. How climate change was mediated to the public through these stage performances revealed the playwrights' – and therefore the audiences' – conflicted relationship with the idea. Each of these four plays grappled with the cultural politics of climate and the unruly boundaries between scientific truth, cultural values, and personal integrity. They each characterised "the climate scientist" in very different ways – the wily science advisor, the youthful enthusiast, the eccentric genius, the detached expert. Ironically, the play that sought hardest to police the truthfulness of climate science – *Greenland*, written by a multi-authored team of writers – was the play that for critics was least successful dramatically (Bottoms, 2012).

Poetry

Ecocriticism emerged in the 1980s as a new identifiable style of literary criticism, and it was followed in the 1990s by a new style of poetry labelled "ecopoetry". For example, the journal *Ecopoetics* was launched in the United States only in 2001. In his *The Song of the Earth*, humanities scholar Jonathan Bate described ecopoetry as "not a description of dwelling with the earth, not a disengaged thinking about it, but an experiencing of it" (Bate, 2000: 42). For Bate, ecopoetry was more phenomenological than it was political. In which case the nineteenth-century English Romantic poet, William Wordsworth (1770–1850), would be the original ecopoet, if not then called out as such. But Bates' view does not capture the diversity of engagements that contemporary ecopoets have with the idea of climate change, not least in their critique of the Romantic pastoral.

Both ecocriticism and ecopoetry had their origins in the environmental activist tradition. Much ecopoetry therefore seeks to raise ecological awareness, including lyrical descriptions of nature often with an elegiac or apologetic tone, especially that of lament. And there is frequent use of mocking or solemn invective to criticise wilful human greed and rapacity. But poets writing about climate change today have developed a much broader range of intentions than merely to deliver "an environmentalist" understanding of climate change. Much poetry inspired by the idea of climate change is now visible in mainstream poetry publishing venues and would not necessarily be contained within the label "ecopoetry" (see Box 7.3). This broadening of poems "about" climate change has been explored by Matthew Griffiths in *The New Poetics of Climate Change: Modernist Aesthetics for a Warming World* (2017). Griffiths deliberately juxtaposes modernist and ecocritical poetics and seeks to expand the modernist canon of climate poetry in unexpected ways. He develops the case that modernist poetic forms are valuable for articulating new and speculative relationships between humanity and nature, older understandings of which have been undermined by the idea of climate change.

In concluding this subsection, I make the suggestion that creative writers are accountable first and foremost to their imagination. Their engagement with the idea of climate change need not – and usually should not – be constrained by scientific knowledge of climate. There are of course competent writers who add scientific data to their works in innovative ways, reflecting the growing collaboration between scientists and artists. One such was the English poet Peter Reading (1946–2011), whose poem "*-273.15*" is a good example of this. By using experimental techniques such as poetic collage, Reading successfully blends the laws of thermodynamics – absolute zero, the lowest temperature physically possible, is -273.15° Celsius – with climatic history.[1] But the general point is that whether a realist, speculative, or Romantic novelist, whether a playwright or a poet, authors are free to create fictional worlds inspired by the idea of climate change but unconstrained by the laws of physics and the methods of science.

Creative writers do not need to follow the climate script of the IPCC. Indeed, it would be suspicious if they did. "While authors have to struggle with the issue of scientific plausibility – this is part of their ethical-professional challenge", says Manjana Milkoreit (2016: 187), "one could also argue that there is no need for climate fiction stories to 'get the science right' ". Literary climate texts are not created to provide science lessons about climate change. This is the danger that Kingsolver's *Flight Behaviour* falls into. Nor do they reveal some universal inner meaning about the significance of climate change, nor instruct readers on how they should respond. The purpose of literary interventions around the idea of climate change is to enlarge and deepen the reader's understanding of its implications for them, *beyond* that offered by the facts of science.

BOX 7.3 CLIMATE POETRY

Throughout this book, I have shown how elusive is the idea of climate change. On the one hand it is beyond human grasping, it is Morton's hyperobject. And yet on the other hand, climate change is also something that is felt at an intense, personal level. Poems can be the ideal literary device to capture and convey such conflicting and disorienting emotions. Perhaps more than other literary genres, poems "can contain open-ended, multiple and even contradictory levels of meaning" (Lidström & Garrard, 2014: 37). In 2019, *Magma* – the UK-based magazine of poetry and writing about poetry – published one of its three editions on the theme of climate change. *Magma* features contemporary poetry that conveys a direct sense of what it is like to live in today's world, whether from little-known or more established poets. By selecting climate change for one of their thematic issues they point to its perceived "ordinariness" for life today. This 2019 themed issue was edited by three guest editors who solicited contributions from multicultural poets in the UK and from poets worldwide. Some of the contributions were speculative science fiction poems, using the medium of poetry to express what climate change means on scales that exceed the present and past, and go beyond ourselves and beyond this time.

The Irish poet Seamus Heaney (1939–2013) won the Nobel Prize for Literature in 1995 and is one of Ireland's most significant poets. In his published collection *District and Circle* (2006), Heaney evokes a powerful contemporary sense of place through his poetic engagements with planetary climate change, environmental risk, and globalisation. One poem from this collection is reproduced in the following, in which Heaney recounts a personal experience of flying above the massive glacier behind the town of Höfn in southeast Iceland.

Höfn

The three-tongued glacier has begun to melt.
What will we do, they ask, when boulder-milt
Comes wallowing across the delta flats
And the miles-deep shag-ice makes its move?
I saw it, ridged and rock-set, from above,
Undead grey-gristed earth-pelt, aeon-scruff,
And feared its coldness that still seemed enough
To iceblock the plane window dimmed with breath,

Deepfreeze the seep of adamantine tilth
And every warm, mouthwatering word of mouth.

Read:

Magma Issue 72: https://magmapoetry.com/archive/magma-72/. See also this anthology of 20 climate poems on the theme 'Keep it in the Ground', curated by Carol Ann Duffy (UK poet laureate 2009–2019): www.theguardian.com/environment/series/keep-it-in-the-ground-a-poem-a-day. Also: Munden, P. (ed.) (2008) *Feeling the Pressure: Poetry and Science of Climate Change*. Berne, Switzerland: British Council.

Visual

Written words allow multiple ways to reimagine the idea of climate change and to expose its meanings through story, plot, film, or poem. The possibilities for reimagining climate change are equally true of the visual image. Visual imagery powerfully shapes the environmental imagination. The famous 1968 "Earthrise" Earth image taken from the Apollo 8 spacecraft has been ascribed agency in the emergence of new forms of planetary awareness and political globalism in the late twentieth century. The essential invisibility of climate change requires some form of rendering by an artist to "make it" visible (Rudiak-Gould, 2013). Artistic practices do just this. This is achieved in one way by climate scientists through their use of climate simulation technologies and advanced computer graphics. Scientists here become artists. Model-simulated future climates are made visual through colourful animations of spinning globes turning into deeper shades of red or purple. Heather Houser (2020) explains how these practices turn climate data into experiential learning – but a learning that is aesthetic and emotional rather than cognitive.

But making climate change visible is also accomplished by other artists using the media of canvas, film, and assorted workable materials. The artist works with the physical manifestations of a changing climate *and* the associative cultural symbols of such changes to represent climate change through materiality, abstraction, and perspective. Visual climate art can accommodate the ambiguities and contradictions of human knowledge and the personal experience of climate change that science sets out to eradicate. The visual artist encourages the viewer to reflect on their own understanding of climate and its changes and, through their own imagination, what this might signify for them.

In her review of ten years of visual climate art, American photojournalist Joanna Nurmis argued that much climate change art has emerged as essentially artistic rather than as propagandistic or activist practice. Art that emerges from these latter practices becomes didactic. Not only does it tend to be dismissed by the art world, says Nurmis, but it also fails to "engage" those who aren't themselves already concerned by climate change. But as with speculative climate fiction, she noted "a strong artistic trend toward the imagery of apocalyptic sublime". This results in art "that may be poignant, but falls out of step with the self-professed motivations of artists and curators alike" to convey concern about the climatic future (Nurmis, 2016: 501).

Visual forms of climate art are displayed in increasing numbers of public, private, or online spaces. These include formal galleries such as the Barbican in London or the Haus der Kulturen der Welt in Berlin, while many artists curate new visual climate content online, with the associated possibilities of animated installations. More informal and opportunistic locations are also used. For example, the feminist artist Juanita Schlaepfer-Miller used an old botanical garden in Zürich to create *Climate Hope Garden*, a participatory art project. This dynamic installation created in material form the effects of climate change on Swiss domestic gardens. Accompanying the installation was a program of workshops for families and school groups, art performances, and talks by botanists, ecologists, and geographers from the Universities of Zürich and Basel. Climate art is also displayed informally as pop-up art in public spaces or alongside climate negotiations. For example, in 2015 the ArtCOP21 climate arts festival took place in Paris alongside the formal UNFCCC negotiations (Sommer & Klöckner, 2021). Hundreds of climate artworks – paintings, collages, photographs, sculptures, videos, installations – were displayed in formal venues such as the Grand Palais and in small galleries, public parks, and cafes.

The power of visual art to explore the interpenetration of climate and culture was creatively exploited in Olafur Eliasson's massive solar installation, *The weather project*, at London's Tate Modern in 2003. In the huge Turbine Hall – the Tate's public gallery – Eliasson created an artificial sun and carefully manufactured an atmosphere of fake weather in which visitors were invited to linger and "dwell" (Figure 7.3). Through this installation, Eliasson was claiming that humans were not just co-creators of *ecological* worlds. Through the confined atmosphere of the Turbine Hall, *The weather project* reminded visitors that humans are unavoidably bound up in the making and experiencing of the weather. But this was happening not just in the Turbine Hall. Human activities are increasingly co-producing the vaster space of the atmosphere and the climates that *it* yields. Through his installation Eliasson was saying that there is no standpoint outside of the weather from which humans can stand and objectively observe, measure, or manipulate the atmosphere (Hulme, 2018). Not even the unconfined atmosphere of the world at large remains immune

Figure 7.3 *The weather project*, Monofrequency lights, projection foil, haze machines, mirror foil, aluminium, scaffolding. Installation view: Tate Modern, London, 2003.
(*Source:* Photo: Andrew Dunkley & Marcus Leith. Courtesy of the artist; neugerriemschneider, Berlin; Tanya Bonakdar Gallery, New York / Los Angeles. © 2003 Olafur Eliasson).

from human touch. The sky is indelibly marked by human hand because people live inside the atmosphere. How can they not? For humans to live culturally with climate is for climate to be inescapably altered.

Photography is a contrasting form of visual representation to that of canvas and brush or material installation. But it is a visual medium that is also circumscribed by the intangible nature of climate change and by the cultural and material influences on its practice. Photography is sometimes understood as offering a technique for faithfully capturing the material traces of objects, landscape, or people. If so, then photographing climate change implies that it must be "there" – it must be somewhere tangible – for it to be captured on film. But photography should not be limited in this way. And neither should climate change. Expecting photographic representations of climate change to be visual conveyors of reliable truth is problematic. The frustrations and paradoxes of photography as a form of representation that makes climate change visible – and therefore "real" – were encountered by Greenpeace International in its 1990s public advocacy campaigns. In her investigation of these efforts, media scholar Julie Doyle revealed the paradox. On the one hand, the photograph is a powerful form of documentation. It carries persuasive power. But photographs cannot escape the limitations of "the visual" as an unmediated representation of reality (Doyle, 2007).

These limitations are especially evident in the case of one particular form of visual image, namely the photographic montage (Mahony, 2016b). Montage is a visual technique that enables creators to import visual formations of distant or imagined places alongside more familiar locations. This creates striking visual juxtapositions. Photomontage produces novel – and sometimes jarring or disturbing – visual renderings of the interconnections of climate change. It is a device that has been widely used to visualise familiar places transformed in the future by the effects of climate change. The flexible possibilities afforded by photomontage offer a visual resource to resolve the "impossible" spatial and temporal scales over which the disruptions of climate change play out. Presumed impacts of climate change can be visualised in what appear to be photographs of future locations but which are in fact overlays of different present realities. Mahony shows that many such photomontages of future climate change revisit Western Romantic imaginaries of "the tropical" or "the ruined". Other montages offer visualisations of climate change–induced migration, problematically reinforcing chauvinistic notions of geographical otherness. As a visual technique of creative representation, montage aims at subversive ambiguity. But by foregrounding the choices, juxtapositions, and arguments that lie behind any representation of climate change, photomontages in fact provoke the viewer into reflecting on their own assumptions of familiarity, difference, and prejudice.

Museums, performance, and practice

Artistic engagements with the idea of climate change through textual and visual forms enable climate change to be reimagined in ways that may be very far removed from "realist" accounts emanating from climate science. The aim of such creative practices through narrative, emotion, and aesthetic encounter is to draw the reader or viewer into different suggestive renderings of the idea of climate change (see Box 7.1). And to provoke audiences into asking difficult questions about climatic knowledge, human responsibility, and the future.

One of the venues where these encounters may most productively occur is the public museum. Alongside the rise in recent years of creative engagement with the idea of climate change has been the rapid growth in the public curation of climate change in museums (Newell et al., 2016). Museums are trusted institutions in the media and political landscape. In many respects they offer ideal venues to give creative form to the idea of climate change through specific collections in local places. In many countries, history museums, art museums, natural history museums, and local museums, museums large and small, are all grappling with how to talk about climate change. But it is difficult to represent climate change in the gallery space of a single museum. There are no self-evidently "climate change objects" to display. Numbers and graphs derived from science do not make for an exciting exhibition and do not appeal to the mixed audiences that museums attract.

Yet museums *can* provide a range of sensorial, affective, and aesthetic experiences for visitors though the agency of art installations, immersive and interactive environments, and audio and other types of performance. Museums can engage in public discussions about climate change, for example through photographs, historic landscape paintings, and textile art that can render climate data in personal and participatory ways. Museums can also allow audiences to investigate how scientific knowledge about climate is produced, represented, and communicated, rather than simply and didactically "convey the facts" (Cameron et al., 2013). These different ways of engaging audiences can activate and broker discussions and interactive dialogues among visiting publics. By harnessing artistic creativities, museums can equip citizens to participate more actively in debates about the policy responses to climate change that will affect their future. In *Curating the Future: Museums, Communities and Climate Change* (2016), Australian museum curator Jenny Newell and colleagues have developed a useful practical guide for displaying climate change in creative ways in museum and gallery spaces. They explore the power of material objects and collections and artistic interventions to stir hearts and minds. And they reflect critically about how and where museums can best intervene creatively to convey the predicament of climate change.

Museums can also offer suitable venues for hosting modes of artistic engagement with climate change that are more directly performative, for example dance and sound. I focus here on sound. Sonification practices and musical performances may activate emotional and experiential responses in the hearer not afforded to the reader of text or the viewer of image. By offering auditory experiences that cultivate aesthetic appreciation, sonic art may diffuse some of the imaginative challenges presented by the idea of climate change – its insensible nature, the abstractions and "distances" of time and space, and the separation of the cultural and the material (Hawkins & Kanngieser, 2017). Two examples illustrate this.

The first is a sonic installation by Scottish artist Katie Paterson, titled *Archive of Vatnajökull (the sound of)*. In 2007, Paterson installed an underwater microphone in Jökulsárlón, an outlet glacial lagoon of Vatnajökull, the most voluminous glacier in the south of Iceland. The microphone was connected to a mobile phone, reachable from any location in the world. She subsequently developed a gallery installation (Figure 7.4) that forged a link between the caller-listener and the real-time movements of the glacial environment. By "dialling up" the glacier, the cracks, groans, and drips of shifting and thawing ice were audible to gallery visitors. Climate change could "be heard". Paterson explained her installation was not only to create a novel audible experience of climate change. It was also to subvert the distancing effect of more standard visual imaginaries of ice melt and glacial ruination where the viewer is detached and remote (Hawkins & Kanngieser, 2017).

The second example is musical composition and performance. In 2015, Dan Crawford, a student at the University of Minnesota, composed a

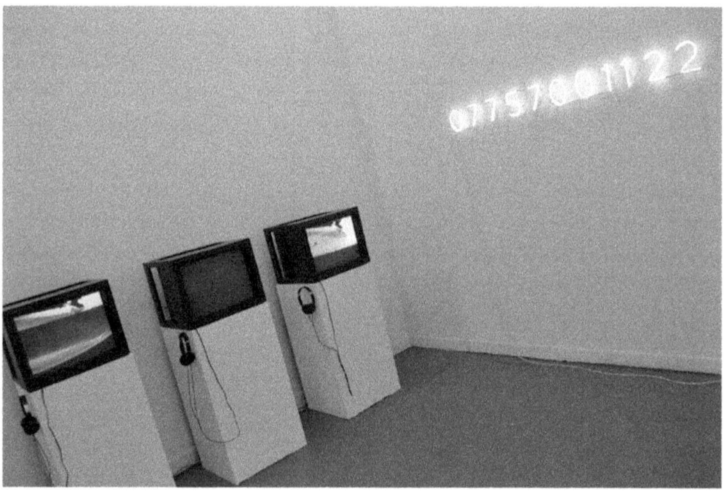

Figure 7.4 Gallery installation *Archive of Vatnajökull (the sound of)* by Katie Paterson. Visitors to the PKM Gallery (Seoul, Korea) could call the number projected on the wall and hear an abstract series of cracks, pops, and creaks as one of Europe's largest glaciers experienced the effects of climate change. During its week-long installation, 10,000 gallery visitors "heard" climate change.
(*Source*: Hawkins & Kanngieser, 2017; reproduced with permission).

song – "Planetary Bands, Warming World" – that traced the rise of northern hemisphere temperatures since the 1880s derived from a standard scientific dataset (Hansman, 2015). The music is performed by a string quartet. Each instrument "played" the temperature range of a different zone of the northern hemisphere, tuned to the average temperature of that region. The cello tracked the equatorial zone, the viola played the mid-latitudes, and the two violins played the high latitudes and Arctic temperatures. Each note of the tune corresponded to a year and the pitch of the note represented the average temperature. Higher notes were warmer years. The audience could "hear" the Earth warming. Music here acts as a creative bridge between logic and emotion, combining logos and pathos.

Chapter summary

The idea of climate change presents profound challenges for creative artists. Not least of these is climate change's abstract nature and the different human subjectivities that shape its meaning. The inhuman distances of time and space between those who cause climate change and those who – in diverse places and times – feel its visceral effects are also difficult for the imagination to embrace. Through combinations of text, drama, vision, installation, sound, and performance, artists confront these challenges to create aesthetic, emotional, and sensory experiences that render climate change imaginable. But they do so not on terms that are laid down by science's explanations of physical climatic phenomena. Artistic inquiry engages with more-than-rational intuitions to provoke the human imagination to reflect on climate change and its multiple meanings. As Tom Griffiths (2007: para. 13) explains for the case of narrative,

> [the story] is a privileged carrier of truth, a way of allowing for multiplicity and complexity at the same time as guaranteeing memorability . . . narrative is not just a means, it is a method, and a rigorous and demanding one. The conventional scientific method separates causes from one another, it isolates each one and tests them individually in turn. Narrative, by contrast, carries multiple causes along together, it enacts connectivity.

So too, I would argue, do poems, visual image, drama, music, and dance. This is why this chapter is offered as a "more-than-science" approach to understanding the idea of climate change.

For the artist, climate change is a mysterious and dangerous idea. It is not one that can be tamed and turned into didactic communicative impulses to "reach" particular audiences in predictable ways. Works of art should not be subordinated to a scientifically prescribed reality. Artists have no obligation to remain "true" to science, although some may seek to work *with* scientists to gain insight and inspiration. Neither need artistic creations seek to instruct or tutor their audience to accept a predetermined view of reality. As critic Robert Boyers explains, "We dislike works of art that are made with too good an opinion of themselves and with an eye to winning the ideological argument of a particular audience" (Boyers, 2019: 40). An artist approaching the idea of climate change should rather seek "to set in motion elements that threaten to escape her control" (ibid). It is therefore not just ecocriticism that should refuse to serve as "the literary department of the IPCC". Visual artists should not become the IPCC's illustrators nor poets the affective conscience of scientists. The artist should not be constrained to deliver creations that "communicate" climate science in any final or assured manner. Good climate art juxtaposes human subjectivities with the epistemic claims that climate science puts into public circulation. It provokes, releases, and celebrates the imagination to explore the multiple meanings of climate change for the human subject.

Note

1 I thank my colleague Chao Xie for this observation and example.

Further Reading

A good starting point for exploring more deeply the questions raised in this chapter about climate change and the social agency of art is **David Maggs** and **John Robinson's** *Sustainability in an Imaginary World: Art and the Question of Agency* (Routledge, 2020). One of the best introductions to the field of ecocriticism is still **Greg Garrard's** *Ecocriticsm: The New Critical Idiom* (2nd edn., Routledge, 2011). Garrard outlines how the relationship between people and the environment is represented through cultural texts. With respect to climate change and literature, there are a number of good reference books to guide you. Two of the best are **Adeline Johns-Putra's** monograph *Climate Change and the Contemporary Novel* (Cambridge University Press, 2019) and **Adam Trexler's** *Anthropocene Fictions: The Novel in a Time of Climate Change* (University of Virginia Press, 2015). **Johns-Putra** also has an edited collection of essays – *Climate and Literature* (Cambridge University Press, 2019) – which takes a wider, multi-authored, and more historical view of how literary cultures have engaged the idea of climate. **David Higgins'** *British Romanticism, Climate Change and the Anthropocene: Writing Tambora* (Palgrave MacMillan, 2017) offers a concise exploration of one particular genre of literature – Romanticism – in relation to climate change. Also recommended is **Guy Abrahams, Kelly Gellatly**, and **Bronwyn Johnson's** *Art + Climate = Change* (Melbourne University Press, 2016), which contains a representative selection of Australian and international artists across 29 climate exhibitions and art events.

Heather Houser's *Infowhelm: Environmental Art and Literature in an Age of Data* (Columbia University Press, 2020) is a valuable exploration of how artists transform the techniques and data of climate science into aesthetic experiences. **Cape Farewell** is one of the longer established climate-art projects and *Burning Ice: Art & Climate Change* (Cornerhouse, 2016) is a reissue of their spectacular 176-page publication comprising 200 stunning colour photographs and illustrations. Its associated website is https://capefarewell.com/. In their book *Curating the Future: Museums, Communities and Climate Change* (Routledge, 2016), **Jenny Newell, Libby Robin**, and **Kirsten Wehner** have brought together a valuable collection of worldwide perspectives on the role of museums in curating climate art.

QUESTIONS FOR CLASS DISCUSSION OR ASSESSMENT

Q7.1: "Ecocriticism is not the literary department of the IPCC" (Garrard et al., 2019: 207). Why would a leading ecocritic who works on climate change fiction feel it necessary to make this claim?
Q7.2: Is Cli-Fi a literary genre?
Q7.3: Amitav Ghosh (2016) has lamented the lack of high-quality creative fiction inspired by the idea of climate change. Consider some of the reasons why this might be the case.
Q7.4: How should artists interpret climate science?
Q7.5: What is the purpose of climate science–art collaborations?
Q7.6: How important do you think museums are for engaging publics with the idea of climate change? Should they focus on communicating facts, proposing solutions, or provoking questions?

8

RELIGIOUS ENGAGEMENTS

. . . Climate change transcended . . .

Introduction

Across the Melanesian archipelago in southwestern Oceania a new genre of climate observatories has recently emerged. Rather than being inspired or instigated by international scientific networks or bilateral aid programmes, they are the creation of the Anglican Church of Melanesia (ACoM). Melanesia possesses a majority Christian population and the Anglican Church has started to use local church infrastructures and communities throughout the region to create a network of scientific observatories. ACoM considers climate change one of the most significant environmental and social issues facing its religious communities scattered across the Melanesian islands. In recent years the Church has forged relationships with local and international organisations to create what they call "ACoM Environment Observatories" (Melanesian Mission, 2019). Using scientific monitoring equipment operated by clergy and lay people, churches are measuring rainfall, storm intensities and durations, and shoreline changes. Daily measurements are transmitted to ACoM headquarters in Honiara on the Solomon Islands where they form the basis for scientific analysis and use by the government.

This innovation by ACoM is attracting the attention of the wider Anglican community. Christian communities in Australia, Samoa, and the UK are developing partnerships with ACoM to extend the environment observatory network. A Green Apostle Award has been established in collaboration with the Melanesian Mission from the UK, the award being made to priests and lay church members who operate observatories across the region. Climate and environmental sciences are being combined with theological and religious education. Bishop Patteson Theological College in Kohimarama on the Solomon Islands offers a curriculum that integrates the study of climate change science with theological training, crossing conventional boundaries of learning between science and religion.

Why is this religious institution mobilising its people and resources to create scientific observatories of climate change? At one level one might see this response as a protective measure. Widespread coastal erosion threatens the well-being and development of many communities in the Solomon Islands. Sea level rise, increased severity of storms and flooding, droughts, saltwater intrusion into freshwater agriculture, and reef habitat loss all threaten to further destabilise local communities. ACoM recognises the need to develop adequate strategies to manage these risks and possible climate-induced relocations. The church proclaims itself as seeking to be good stewards of the Solomon Islands for future generations. At another level one might see these environmental observatories as an institutional response to a deficient national infrastructure. The Solomon Islands government does not have sufficient systems in place to monitor ongoing climatic and environmental change. As a prominent cultural institution, ACoM considers this an opportunity. Rather than import costly monitoring equipment and expertise from abroad, the observatories repurpose existing church infrastructure and personnel. Future plans include expanding observatories to all islands hosting ACoM churches and integrating scientific observing alongside the priestly duties of clergy.

Do these church-run observatories reveal a "science-based" approach to understanding climate change? Or is there a larger narrative of meaning and acting in the world into which ACoM is appropriating science? In his study of ACoM, geographer Adam Bobette suggests there is. The Solomon Islands have been Christianised for over a century and the Anglican Church is powerfully embedded in the cultural imaginary of these communities. Rather than seeing climate change as a "secular scientific fact", Bobette shows how for Solomon islanders the idea of climate change is assimilated into the powerful gravitational force of an Anglican cosmic imaginary. "Climate change, as a narrative about the natural world", says Bobbette (2019: 558), "could not be disentangled from an Anglican narrative of nature". Understanding climate change in the Solomon Islands therefore requires understanding the histories and geographies of religious conversion. After embracing the idea of climate change according to its own theologies, myths, and materialities, it seemed self-evident for ACoM to exercise its cultural status in the islands by serving as a platform for scientific observation. As Bobette observes, "The imagined divide between climate change scientists and Christian reactionaries is being undone by Christian communities actively engaging in making sense of climate change through an Anglican cosmic framework" (p. 557).

There is no religiously inspired climate denial here. Nor do Melanesian Anglicans view themselves as helpless victims of forces from outside. They are active in describing what is happening to their local environments, whilst at the same time making climate change meaningful within their religious imagination and everyday practices and duties.

***

Many communities and individuals around the world interpret the idea of climate change according to specifically religious frames of reference (Jenkins et al., 2018; Berry, 2022). These religious beliefs, narratives, subjectivities, and rituals filter and interpret different knowledge claims about climate: *What* is happening and why. They also shape individual and communal ethical and social behaviours: *Who* should act and why. Faith communities therefore offer what I call – following the anthropologist Clifford Geertz – "thick" accounts of moral and normative reasoning that guide action in the world (Hulme, 2017). Such religiously inspired responses to climate change are in contrast to many secular calls for climate mitigation and adaptation policies that rest on "thin" global values. Secular values may be widely communicated, but they are often culturally nonspecific and may be perceived to offer shallow moral criteria for motivating and guiding ethical action.

The chapter starts from the premise that the "comprehensive doctrines" offered by religions – the integration of religious thought, practice, duty, and identity – remain vitally important for many in today's world as they seek to make sense of the human experience (Dwyer, 2016). On the other hand, the mere fact of religious engagement with climate change hardly resolves the deep divisions, tensions, and dilemmas within the social world that the idea provokes. It may do just the opposite. The 2014 Interfaith Statement on Climate Change from the World Council of Churches (Interfaith Summit, 2014) and Pope Francis' 2015 Encyclical *Laudato Si': On Care for Our Common Home* (see Box 8.2 for further elaboration) offered hopeful visions for common action on climate change. But there remain too many tensions between and within different religious traditions for this hope to be easily realised. Yes, the world's religions have increasingly engaged with the idea of climate change. But they do so for different reasons and in different ways. Faith traditions all engage with the scientific account of climate change, some more sceptically and some, as in the example of the Anglican Church of Melanesia, more constructively. But what climate change means for religious communities is driven primarily by their shared beliefs, sense of moral duty, social solidarities, and individual subjectivities.

This chapter outlines varieties of religious engagement with the idea of climate change that have emerged over the last three decades. It is organised in three main subsections. First, I consider religious systems of thought and meaning through the three lenses of cosmologies, eco-theologies, and religious mythologies. I then give consideration to how these systems of thought applied to climate change motivate and orient ethical action in the world. The third main subsection considers the relevance of some sociological aspects of religions, namely rituals, identities, and institutions. The chapter draws upon **religious** and **cultural geographies, sociologies of religion**, and insights from **eco-theology** and is illustrated through different religious traditions, for example Roman

Catholicism, American evangelicalism, Islam, Confucianism, and Buddhism. The overall argument of the chapter is that religious faith and practice offers a "more-than-science" approach to making sense of the idea of climate change.

Religious thought

Religious cosmologies and mythologies shape the lifeworlds of believers and – indirectly but still importantly – those of many unbelievers (Figure 8.1). These cosmologies influence how people make sense of unsettling changes in their local climatic environments. Systems of religious thought lend substance and power to social and ethical norms, enhance social capital, and valorise certain lifestyles whilst admonishing others. Many commentators have argued that climate policies need to tap into intrinsic, deeply held values and motives if cultural innovation and change is to be lasting and effective. Without understanding the religious and spiritual dimensions of peoples' subjective experience of the world, climate change communication, advocacy campaigns, and climate policy development and implementation will likely be deficient. Articulating these different cosmological accounts of climatic agency and blame – and giving them salience – is a task of religious studies scholars. These frameworks of religious thought and belief offer interpretative resources for human being, knowing, and acting in the world. They are quite different to the resources put into public circulation by science. The subsections that follow examine aspects of religious thought in relation to climate change using the categories of cosmology, eco-theology, and religious mythology.

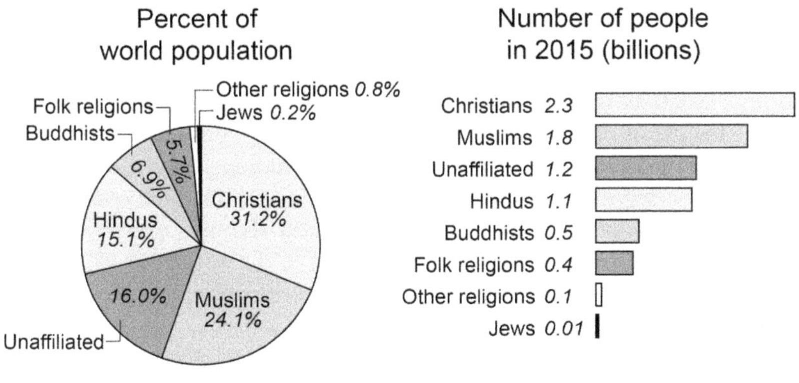

Figure 8.1 Distribution of the world's population in 2015 according to religious affiliation (including non-affiliates).
(Source: Redrawn from Pew Research Center).

Cosmologies

In his influential 1967 essay titled "The historical roots of our ecological crisis", the historian Lynn White Jr. observed that "What people do about their ecology depends on what they think about themselves in relation to things around them. (It) is deeply conditioned by our beliefs about nature and destiny . . . that is by religion" (White, 1967: 1205). Although his wider thesis that blame for ecological damage can be laid firmly at the door of Christian theology has been largely discredited, he was right about the significance of religious worldviews for influencing people's actions in the world.

One of the most significant functions ascribed to religion is its creation and defence of a cosmology. Cosmologies are imaginative systems of thought – often expressed in the form of stories and myths – that communicate fundamental assumptions about the origins and nature of the cosmos and the ultimate purpose of human existence. Cosmologies offer to explain the ordering of the natural world, the divine, animate and inanimate agencies at work within it, and the appropriate duties and responsibilities of human beings. They may also point towards exemplars of good, meaningful, and worthwhile lives. So it is unsurprising that people's cosmologies will influence what they think about climate change. Scientific materialism may be said to offer one such (nonreligious) cosmological narrative. But scientific claims that humans are largely responsible for changes to the world's climate may be unconvincing if they do not engage with – or make sense of – the spiritual and moral lifeworlds people inhabit.

Cosmological accounts of blame for climatic disturbances come in many different guises. Among Marshall Islanders in western Oceania, for example, the blending of Indigenous and Christian cosmologies offers accounts of climatic agency and blame very different from those that would be understood in most Western settings (Rudiak-Gould, 2012). In other cosmologies, the effects of weather and climate on local ecologies and physical landforms – such as rivers, forests, glaciers – are understood to be mediated by the degree of respect people show through their behaviour for spiritual beings who may inhabit such dwelling places. Although not always fitting the formal category of "religious beliefs" (Leduc, 2010), such cosmologies nevertheless challenge modernist and materialist accounts of why climates change. For example, for some Hindus, belief in the epoch *Kali yuga* allows the divinity of the sacred Ganges River to offer reassurance in times of climatic disturbance.

A cosmological approach for understanding climate change can unsettle the markers of rationality that are conventionally assumed to distinguish early modern from late-modern cultures (see Chapter 1). For example, Barnett (2015) shows how early modern scholars deemed the Biblical Flood "anthropogenic" in the sense that it was divine punishment induced by human sin. But this is not so far from the contemporary belief expressed by the eco-theologian Michael Northcott when he proclaims that "Global warming is the Earth's judgement on the global market Empire and the heedless consumption it fosters" (Northcott, 2007: 7). In either case, the cause of global climatic change – whether mediated by God or by "the Earth" – is the moral failures of humanity.

Eco-Theologies

Theology emerged as a discipline of thought in fourteenth-century Christianity and is most commonly associated with the Abrahamic religions of Christianity, Judaism, and Islam. "Eco-theology" however is a subspecies of theology that refers "to the construction of religious doctrines and teachings to emphasise the appropriate human relationship to nature" (Dalton, 2018: 271). Virtually all religions possess such teachings (Deane-Drummond, 2008). The articulation of eco-theologies within different faith traditions was accelerated through a series of ten conferences on religion and ecology convened between 1996 and 1998 by the Harvard Center for the Study of World Religions (Grim & Tucker, 2014). These conferences brought together religious scholars of Buddhism, Christianity, Confucianism, Hinduism, Indigenous peoples' spirituality, Islam, Jainism, Judaism, Shinto, and Taoism, with a view to re-examine the scriptures, doctrines, and rituals of these faiths with respect to humanity's place within creation. Significant statements about climate change were issued from this process by the spiritual leaders of the Eastern Orthodox Church (Bartholomew I, the ecumenical patriarch of Constantinople) and Buddhism (the Dalai Lama).

Eco-theologies offer systematic and analytical frameworks and resources for thinking about climate change, for diagnosing the nature of the phenomenon, and for pointing towards its possible remediation. Eco-theologies help the religious think about "boundaries": Between the divine and the created, between the natural and the artificial, and between the human and non-human. For example, the Hindu emphasis on interconnectedness and moral solidarity with all things is suggestive of a stance towards climate change that resists reducing it to merely a technical question of greenhouse gases and energy technologies. Eco-theologies also help the religious relate climate change to other questions of justice, for example those of social justice raised by the liberation theology of the Christian church or *Maslahah* in Islam, the concept of public interest and welfare. Eco-theologies also have bearing on religious eschatology – the study of "end-times".

Eco-theologies inform confessional statements or faith creeds of why a particular religious tradition thinks climate change is a moral issue – or at least what *sort* of moral issue climate change is. Such statements can inspire political activism or guide the behaviours of the faithful. One example of a confessional statement on climate change was issued by Operation Noah, a Christian evangelical coalition based in the UK. In February 2012 they issued a public statement on climate change, the so-called Ash Wednesday Declaration (Operation Noah, 2012). It challenged the Christian church to recognise that care for God's creation – and therefore concern about climate change – was foundational to the Christian gospel. This declaration deliberately echoed the 1934 Barmen Declaration, which first gave coherence and visibility to the emergent Confessing Church, which resisted the Nazi regime. The subordination of the church to the Nazi state was seen as a confessional issue, an issue that touched the very heart of faith, not something that German Christians could politely agree to disagree about. By association, Operation Noah claimed climate change to be just such another "confessional issue". "For our generation", they declared, "reducing our dependence on fossil fuels has become essential to Christian discipleship".

Religious mythologies

Mythologies are potent and enduring cultural inventions. In essence, myths may be thought of as powerful stories about something significant. They transcend the rational category of truth that logic and science rely on. Rather, myths engage the notion of truth at a deeper level of meaning and thereby exert a compelling influence over the human imagination. Myths give expression to non-reductive aspects of the human experience of the world – such as love, hope, desire, fear, tragedy – that can only be captured through imaginative and compelling stories. Because they acknowledge the reality of human desire and the possibility of transcendent hope, myths are indispensable for helping many people live with the hardship, uncertainty, and ambiguity of life. Simply put, mythology extends the scope of human beings. For religious scholar Karen Armstrong, it is in *this* sense that science and myths should be thought of as similar. Neither science nor myth is "about opting out of this world", she says, "but [about] enabling us to live more intensely within it" (Armstrong, 2005: 3). Science and myth offer us different ways of achieving this end.

The philosopher Lisa Stenmark (2015), in her reflections on mythology and climate change, makes a further important distinction: Between theology and religion. *Theology* she argues – and to this might be added science – concerns itself primarily with doctrinal statements, declarations of what is believed to be the case, i.e., what is deemed rationally true. Which helps us to see that in *this* sense both theology and science are opposed to myth and storytelling. Myths, as with fictional stories, are not to be taken literally. A myth is "true" not because it conveys factual information but because it is effective (Armstrong, 2005). It is performative. If a myth gives new hope and direction to a community or revitalises and repurposes a person's life, then it is valid.

For Stenmark, myth-making is therefore a religious *art*, distinct from theology. Rather than issuing truth-claims through logical reasoning, the goal of religious myths is to provoke and inspire. Myths therefore offer ways of organising human thought and practice about climate change in ways quite different than those offered by confessional statements or doctrines drawn from theological reflection. The role of myths is to increase the plurality of perspectives on the world, to open closed minds to alternative ways of thinking. It is to point to absolutes when people fail to act and to point to uncertainties when people become too confident. We considered some of these functions of myth through the case of "the trickster" in Chapter 6 (see Box 6.1).

Creating and curating myths is an art that the world's great religious traditions have long mastered, as suggested by the Alliance of Religions and Conservation (ARC) in their climate change statement (ARC, 2007):

> The emphasis on consumption, economics and policy usually fails to engage people at any deep level because it does not address the narrative, the mythological, the metaphorical or the existence of memories of past disasters and the way out. The faiths are the holders of these areas and without them, [climate] policies will have very few real roots.

Religious mythologies offer stories that help people to judge and act in the midst of uncertainty, not least with respect to the troubling disorientations introduced by climate change. Religious myths therefore become cultural, imaginative, and even political, resources that can be called upon in times of crisis, transition, or confusion (see Box 8.1 for the example of Noah's Flood). Myths should not be seen as a barrier to a "correct" scientific understanding of climate change. Stenmark recommends embracing the ambiguity, complexity, and partiality of religious myths as a means of escaping the illusion that science will ever yield all that is necessary to know about the future to adequately guide actions in the present. Religious myths offer powerful and enduring narratives of meaning that bind people together in a shared experience of the world and of how they (should) relate to each other.

BOX 8.1 THREE INTERPRETATIONS OF NOAH'S FLOOD

For Middle Eastern and Western cultures, the flood myth is the archetypical account of existential threat brought on by climatic disaster. The Biblical version of the Noahian Flood was likely written down late in the second millennium BCE. The unprecedented and divinely ordained rainfall led to the destruction of all sentient life, barring Noah, his family, and breeding pairs of creatures who were all saved through God's mercy. A similar outcome occurs in the Sumerian flood myth believed to date from the fourth millennium BCE, humanity only being rescued by the boat-building exploits of the philosopher-king Atrahasis. This flood myth continues to be evoked in many popular fictional writings and film scripts (Salvador & Norton, 2011), such as J.G. Ballard's novel *The Drowned World* (1962) and recent climate change movies such as *The Day After Tomorrow* (2004) and *The Age of Stupid* (2009).

Geographer Hannah Fair has followed the story of Noah's Flood to the Oceanic island nation of Vanuatu. Fair invokes the idea of *tufala save* (literally "double knowledge") – the balancing of multiple epistemologies of climate change – to explore the meanings of this myth in Ni-Vanuatu and wider Oceanic cultures (Fair, 2018). Fair shows how the Noahian myth manifests in three different ways within these islands. In one reading, the rainbow is a signifier of God's covenant promise never again to flood the Earth and hence becomes a reason for scepticism about the reality of climate change. God would never allow the land to flood. But the myth is read differently by others in Vanuatu. For some, Noah embodies the virtue of preparedness and thus inspires appropriate pre-emptive and adaptive responses to a rising ocean in these low-lying islands. Fair also found in circulation a third reading of the myth. By focusing on those who had been left *outside* the ark, an exclusion they would claim was unjust, some church leaders and climate activists find inspiration from Noah's Flood for a justice-based narrative of climate change.

Fair's study highlights "the potential for more-than-scientific yet not anti-scientific responses to climate change that are locally meaningful and morally compelling" (p. 1). Rather than "the rainbow promise" only inspiring scepticism about climate change, Fair shows how a religious myth can provoke radically different political responses to the idea of climate change. Different meanings of climate change are articulated through different interpretations of Noah's Flood, each of which navigates between different types of knowledge (*tufala save*), between religious myth and Western science. The case study illustrates the distinctly geographical project of rendering climate change locally meaningful, in this case through Christian story-telling. The broader lesson is that religious myths offer rich interpretative resources for people to make sense of climate change on their own terms.

Read:

Fair, H. (2018) Three stories of Noah: navigating religious climate change narratives in the Pacific Island region. *Geo: Geography and Environment*. 5(2): e00068. Also: Salvador, M. and Norton, T. (2011) The flood myth in the age of global climate change. *Environmental Communication*. 5(1): 45–61.

This way of approaching the idea of climate change is one also favoured by author and climate advocate Alex Evans in his short book *The Myth Gap* (2017). Evans wrote this book after 20 years' experience working inside the international political networks and institutions that broker deals and enact policies on climate change. But in *The Myth Gap*, Evans calls for a rediscovery of myth within Westernised, secular societies. Decarbonised futures for humanity – such as imagined in the 2015 Paris Agreement – cannot be ends in themselves. Such futures, he argues, only gain strength and legitimacy by being placed within larger narratives about what guides and energises people to act in the world, within cultural accounts of meaning and purpose. For Evans, this is what can be offered by religious myths. The danger of too narrow and instrumental a focus on achieving "low carbon societies" or "net-zero emissions" is that this focus is reductive in vision and weak in moral force. It fails to deal with ultimate questions of the good, the just, the meaningful, the transcendent.

These are ultimate philosophical questions that any utopian project has to grapple with and they are ones to which religious mythologies can give coherent and meaningful answers. They are also questions that religions do not shy away from engaging with: What is it to be *imago dei*, to be made in the image of God? What does it mean for people to be emancipated? What ultimately is a satisfactory human life? These are exactly the questions that drove Pope Francis' (2015) encyclical on climate change, *Laudato Si'* (see Box 8.2).

Ethical responsibilities

For the reasons summarized earlier, religious faith traditions and communities develop "thick" accounts of moral action in the world – in response to climate change as much as to other social and ecological challenges. One example of such an account would be the idea of the Kingdom of Heaven on Earth. This is the idea that people made in the image of God are called on and inspired to bring about justice, goodness, and freedom for all, a vision that has long been part of many religious movements. Or another would be the Buddhist-inspired Bhutanese tradition of *chos srid gnyis* (literally "the dual system"). This underpins Bhutan's commitment to a system of governance that promotes Gross National Happiness (rather than Gross Domestic Product) through a unity of religion (*chos*) and politics (*srid*).

Such religious accounts of moral guidance and inspiration contrast with many secular calls for climate mitigation and adaptation that rely upon "thin" global values. Thin global values may offer ethical criteria that may be widely acknowledged intellectually but that lack conviction. They may often be abstract, rootless, and culturally nonspecific (Wolf & Moser, 2011). Thin global values do not capture the full range of concerns and commitments expressed by local communities affected by a changing climate, for example, their concern for sacred landscapes and commitment to moral integrity, social solidarity, and justice (Kerber, 2010). They do not breathe life into the motivational moral commitments and energies that can enact and guide personal and social change. These moral commitments are what Jürgen Habermas refers to as what is "missing" in secular societies (Habermas et al., 2010) and, in similar vein, what Naomi Klein describes as the "full moral voice" that is missing in the climate movement (Klein, 2014: 464).

In contrast, by drawing upon cosmological accounts of reality – and given coherence and emotive expression through eco-theologies and mythologies – religious faiths can offer "thick" accounts of moral reasons for guiding ethical behaviour. Any attempt to understand how religions engage with the idea of climate change must therefore include an appreciation of how different faith traditions reason ethically. In relation to climate change this would include reflection on – as a minimum – human responsibility to the non-human world, to the human other and with respect to the future. From a Christian theological perspective, these responsibilities might be captured through the notions of "creation care", "neighbour care", and eschatology (Wilkinson, 2012). Different religious faiths might frame these responsibilities slightly differently. And within any single faith tradition there will be different interpretative positions on how they should be discharged. One of the more comprehensive statements has come from the Catholic Church through Pope Francis' (2015) encyclical (see Box 8.2 for elaboration). But however religions engage these questions of human duty and responsibility, they are ones that secular economists are also well familiar with. They lie at the heart of the problems of economic valuation and discounting (see Chapter 3): What relative value should be placed on the non-human world, human others, and unborn future generations?

BOX 8.2 *LAUDATO SI'* – POPE FRANCIS' 2015 ENCYCLICAL

In June 2015, Pope Francis issued an encyclical of the Roman Catholic Church titled *Laudato Si': On Care for Our Common Home*. An encyclical – or papal letter – is a teaching document at nearly the highest level of doctrine in the Catholic Church. It is meant to be carefully considered by all believers with respect to their everyday lives and the life of the church. Pope Francis directed his encyclical to "every person living on this planet" (§3), not only Catholics or even just Christians. He wanted people of other faiths – and none – to also consider its message and to translate it into their own traditions and terminology. *Laudato Si'* – meaning "Praise be to you" – is taken from a canticle of the thirteenth-century Saint Francis of Assisi, an *Common Home* therefore appropriate source of inspiration for the Argentinian Pope who deliberately took upon himself the name of Francis after commencing the Papacy in 2013. He was thereby deliberately drawing attention to the long tradition of Catholic teaching and practice about respect for the natural world.

The encyclical was an important intervention in the public politics of climate change from a major religious leader, perhaps the world's most visible religious leader. Its significance lay not so much in his comments about the reality of climate change. Nor about its call to limit global temperature to 2°C of warming. (In fact, the encyclical says nothing about this policy goal.) What was most significant was that he placed the idea of climate change within a much larger story, one about what it means to be human, what it means to be made in the image of God. Pope Francis was concerned first and foremost with offering a vision of human dignity, responsibility, and purpose, drawing upon the rich traditions of Catholic theology and ethics.

On Care for Our Common Home therefore offers a powerful story – a metanarrative – about the human condition. Concerns about technology, water, power, climate, slavery, biodiversity, and greed are woven together into an inspirational account of divine goodness and healthy human living. The encyclical escapes the confines of a narrowly drawn science and economics and shows the power, vitality, and inspiration of a Christian worldview. It also draws attention to the centrality in the Christian faith of the idea of the transformation of the human person: "For this reason, the ecological crisis is also a summons to profound interior conversion . . . an 'ecological conversion', whereby the effects of their encounter with Jesus Christ become evident in their relationship with the world around them" (§217).

(Continued)

Laudato Si' offers a comprehensive account of the human condition that embraces science but is not driven by science. It places concerns about climate change, the integrity of ecosystems, and social justice within a much more capacious and integrated worldview. It is rooted in a cosmology that has both material and spiritual dimensions and that recognises the human capacity for ingenuity but the propensity for greed. *Laudato Si'* provides an interpretative framework for the faithful – and for "all people" – to respond ethically to the human challenges revealed by climate change.

Read:

Pope Francis (2015) *Praise Be to You – Laudato Si'. On Care for Our Common Home*. San Francisco: Ignatius Press. See also the collection of essays in the 2015 Special Issue of the journal *Environment: Science and Policy for Sustainable Development*. 57(6).

Within many faith traditions the idea of stewardship – sometimes "creation care" – or *Khilafa* in Islam – has been of central importance when approaching the question of climate change. The idea of environmental stewardship develops an account of ethical responsibility for humans as "trustees of the planet". The planet – and by association its climate – is God's true property and yet people have been entrusted by God to exercise due care and diligence over its management. Stewardship as an environmental ethic has also been adopted by some climate scientists. For example, the American National Oceanographic and Atmospheric Administration runs a public educational programme called "Planet Stewards". Yet stewardship is contested within some eco-theologies if it implies a pre-eminence of the human over the non-human. Many Indigenous faith communities also adhere to notions of "trusteeship" with respect to climate. In these cases, it is often a stewardship to be exercised as members of a shared ecological community – humans as well as animals, plants, and soils – rather than in obedience to – or worship of – a transcendent deity.

Another line of ethical reasoning present within numerous religious traditions and that is relevant for thinking ethically about climate change is that of virtue. Virtue (character-based) ethics has taken its place in secular thought alongside **deontological** (duty or rule-based) and **utilitarian** (consequentialist-based) ethics. All three modes of ethical reasoning may be helpful in different ways for thinking about human responsibility and climate change – virtues, rules, and consequences are all important for thinking ethically about actions. For example, when justifying actions that seek to reduce the dangers of climatic hazards, these contrasting ethical traditions might argue as follows. A virtue ethicist would point to the fact that doing so would be charitable or benevolent; a deontologist to the fact that doing so would be acting in accordance with a moral rule such as "Do unto others as you would be done by"; a utilitarian to the fact that the consequences of doing so will maximize well-being.

I emphasise here virtue ethics because of its endorsement within many religious traditions. Although the Western tradition of virtue – the idea of moral excellence – originates in Classical Greek philosophy, it has long been valorised within the Christian, Jewish, and Islamic traditions. For example, the thirteenth-century Dominican Friar Thomas Aquinas promoted the virtuous life as one in which "good habits bear on activities". There is also a long history of virtue ethics in eastern religions, notably Buddhism and Confucianism. For example, within Confucianism is the idea of *ren*, the virtue that characterises a human being as a *human* being. The moral sense of *ren* – and its relevance as an environmental ethic – is captured by the Buddhist idea of "ten thousand things". The virtuous person "takes heaven, earth and the ten thousand things all as one body, no part of which is not oneself" (Huang, 2017: 53). A virtuous Confucian takes care of the bird, the stream, the tree, the air, not because of their intrinsic value. They do so because they are part of their own body. The essence of virtue, then, is its emphasis on the connection between goals and practice, between recognising the person one believes oneself to be – or wishes to be – and the means of achieving that vision.

More so than deontological or utilitarian ethics, virtue is an ethic rooted in specific places, histories, and practices (MacIntyre, 1981). It is nurtured in lived communities rather than arrived at through abstract reasoning (deontology) or consonant with liberal individualism (utilitarianism). Virtue has both individual and civic expressions, which emerge from collective moral reasoning. Civic virtue arises from the cultivation of individual habits and behaviours that can transform the success of a community. Practicing virtue can therefore offer both instrumental and normative benefits for dealing with the challenges of climate change (Hulme, 2014a). *Instrumentally*, when contributing to public discourse and deliberation about responses to climate change, the exercise of virtue can inspire and deliver change. And, *normatively*, it is recognised that striving for moral excellence is virtuous: Virtue is indeed the true goal – the telos – of being human. For example, Forrest Clingerman and colleagues applied the ethic of virtue to the question of whether or not the world's climate should be deliberately engineered through interventionist technologies, so-called solar geoengineering (see Chapter 10). The qualities of character they suggested would be needed to pursue such a project were the virtues of responsibility, humility, and justice. They illustrated these qualities with reference to the traditions of virtue within Buddhism, Islam, and Christianity (Clingermann et al., 2017).

Religious sociologies

It is not simply in terms of cosmologies, confessional creeds, or myths that religions are relevant for understanding how the idea of climate change engages the human mind. Nor are religions important only for upholding certain moral principles or offering ethical guidelines or directives for individual and collective behaviour with respect to climate. Just as important for understanding climate change in religious terms are aspects of the sociology and anthropology of religions, for example religious *rituals*, *identities*, and *institutions* that shape community and individual behaviours. Cultural geographers such as Claire Dwyer draw attention to the importance of religious faith as a legitimate social identity, even if sometimes overlooked by secular sociologists (Dwyer, 2016). She argues that understanding everyday "lived" religion – through material, embodied practices and the visual performance of faith (i.e., rituals) – reveals how people connect the mundane with the transcendent. Religious practices can not only ameliorate hardships, including climatic hazards, affecting individual and communal life. They can also inspire alternative value systems and lifestyles that have bearing on climate change. The following three subsections look briefly at these three sociological aspects of religions in relation to climate change.

Rituals and sacred nature

Rituals fulfil a very important function in forming and sustaining religious identity and solidarity within communities of faith. Rituals are usually engaged with collectively and cultivate a sense of awe, fascination, or dread with respect to the sacred. They also seek to mark – and to maintain – the understood boundaries between the sacred and profane. Rituals often act-out religious myths through stylised material symbolism and are a central part of the experience of both organised (centralised) and less organised (local) religions. Religious rituals have long been performed in relation to the human experience of weather and climate and for many religious communities such rituals continue today. These include rituals of thankfulness for a good harvest secured – traditionally linked to the behaviour of the weather – and rituals of supplication to the gods in order to secure a good harvest (Hardwick & Stephens, 2020). Religious rituals can hold meaning and importance for those who might nevertheless doubt their material efficacy. They act as a mark of identity and belonging. When climatic or weather disaster strikes, communal prayers and processions can strengthen a community's solidarity and common identity in the face of hardship.

Religious rituals and festivals are an essential part of the "thick" moral discourses alluded to earlier. They perform particular understandings of how a faith community understands its relationship to the physical world around them, in the case considered here in relation to weather and climate. For example, the pilgrimage to the village of Santuario de Chimayo in the USA attracts 300,000 Catholic pilgrims each year to this small village in northern New Mexico, famous for its miraculous "holy dirt". This mass walk cultivates a sense of profound respect amongst its participants for the Earth and the natural world. Or in Peru, the ancient Festival of Quyllurit'I – literally in Quechua "Lord of star (brilliant) snow" – pays homage to a miraculous apparition of Jesus that occurred near the Ausangate glacier. It attracts 10,000 pilgrims annually and offers a powerful expression of respect for and solidarity with Peru's threatened mountain glaciers. It is through rituals and festivals such as these that many religious communities form their sense of connection to the natural world, including the climate. For the idea of climate change to resonate with such religious traditions it needs to encompass much more than a scientific description of what is happening physically in the atmosphere.

Religious identities

A focus on religious identity has been important for many scholars seeking to understand the motivational dimensions of social responses to climate change. One common analytical approach to achieve such an ambition has been the use of survey instruments to segment populations according to people's self-declared religious identity. Such studies have sought to reveal how different religious affiliations and identities correlate with a variety of climate change beliefs, attitudes, and behaviours. This work has been most extensively conducted in the United States, but comparable surveys have been conducted in several other countries (Koster & Conradie, 2019). In the USA, evangelical Protestant Christians seem much less likely to affirm the reality of climate change and to attribute it to human actions than do other religious identities. American evangelicals are therefore much less personally concerned about climate change than the average American citizen (Figure 8.2). Hispanic Catholics on the other hand are much *more* concerned. How far these differences of expressed beliefs and attitudes at a population level in the USA are a function of religious identity as opposed to coincidental political identities is not entirely clear (Veldman, 2019).

There are other less reductive ways of thinking about religious identity in the context of individual engagement and collective action around climate change. For example, in a study of church community engagement with climate change in Scotland, Kidwell et al. (2018) found evidence of what they termed "eco-theo-citizenship". The Eco-Congregation religious groups these authors studied were not driven primarily by "the issue of climate change". Rather, what mattered to them was the benefit to "community-building"

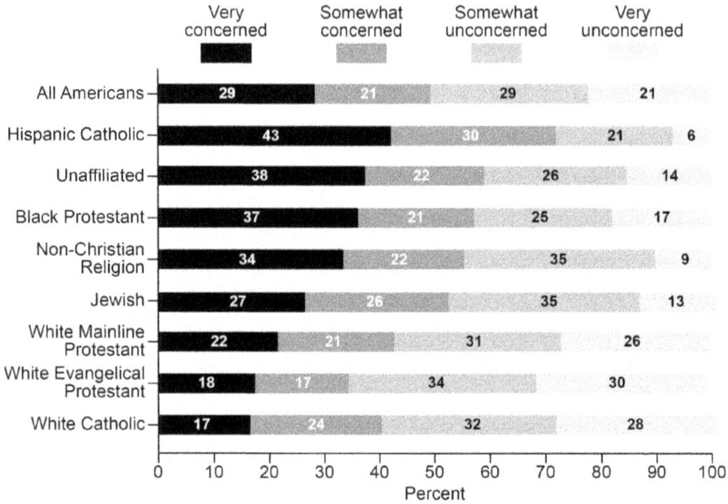

Figure 8.2 "Climate change concern" in the United States, segmented by self-declared religious affiliation.
(*Source*: Redrawn from Jones et al., 2014).

activities that engaging with climate change might bring. They sought to nurture an identity of religious environmental citizenship that spanned multiple political scales from local to international. Primary values emphasised by these groups included "environmental justice" and "stewardship". Pro-environmental behaviours, theological beliefs, and citizenship were combined in a mutually reinforcing spiral, which cultivated a sense of social solidarity and community participation. Concern for climate change was subservient to the more important religious impulse of these Eco-Congregations to express their religious identities through civic action and personal duty.

Religious institutions

Beyond shaping rituals and individual identities, many organised religious faiths maintain an active presence in the world through their exercise of institutional power. This form of power can be of great significance for the engagement of religion with climate change. Religious faiths such as Christianity, Islam, and Hinduism possess substantial institutional and economic resources, as well as possessing significant political power in certain jurisdictions. As with other nonstate actors in the climate governance regime – such as businesses, cities, and NGOs – religious movements and institutions have the mobilizing power to enlist and de-list multitudes of citizens in influential causes. This is particularly the case for religions with more centralised and hierarchical structures, such as Roman Catholicism and Shia Islam. Religious actors become key contributors to climate political discourses at local, national, and international levels. Climate activists regularly cite the importance for international climate negotiations of the presence of faith groups (see Chapter 9). And influential climate scientists have publicly called for enhanced collaboration among religious institutions, policy-makers, and the scientific community (Dasgupta & Ramanathan, 2014).

Religious institutions have always played an important role in public political life for large parts of the world. The presumption of religious decline and the secularisation of society – a narrative that gained tenacious hold amongst Western intellectuals during the twentieth century – has proven false. During the last quarter century, religious institutions, identities, and expressions of public religiosity have been resurgent in many Western societies. Rather than being described as "post-religious", Western societies are now better described as "post-secular". Worldwide this resurgence of religion has given rise to what religious studies scholar Erin Wilson calls "new religious globalisms" (Wilson, 2012). Wilson shows how these religious globalisms are important, powerful, and creative in relation to public policy issues concerning human rights, aid and development, conflict resolution, and, not least, climate change. Neoliberal religious globalisms, religious justice globalisms, or neo-traditional religious globalisms all position themselves in relation to climate change in different ways. But however manifest, they certainly cannot be ignored. Wilson's thesis is that the forms of social identity and political power refracted through religious institutions present a challenge to the Western assumptions of secularism inherited from the twentieth century. The emergence of new religious globalisms – and the institutional power that lies behind them – offers new narratives on climate change beyond those emanating from Western secularism.

Religious institutions are also important at the local scale. In some cultures and in some regions, religious institutions may hold more social – and even economic – capital than other civic actors. They act as the primary conduits through which people and resources are mobilised to address some of the challenges of climate mitigation and adaptation. Religious institutions have even been involved in climate observation and monitoring, as seen earlier in the case of the Anglican Church of Melanesia.

Indonesia is a good example of this. Islamic institutions in this country play significant roles in leading public responses to climate change (Koehrsen, 2021). The two largest civil society organisations in the country – Muhammadiyah and Nahdlatul Ulama – both mobilise their institutional resources to tackle local challenges presented by a changing climate. Over 75 per cent of Indonesia's more than 200 million Muslims are affiliated to one or other of these two organisations, which are able to call upon schools, universities, mosques, *pesantrens* ("Islamic boarding schools"), and *pengajians* ("Islamic teaching forums") as religious-cultural resources. Although the Muhammadiyah advocates a more modernist expression of Islam and Nahdlatul Ulama a more traditional stance, both organisations promote *Jihad bi'iyah* ("eco-jihad") as a way of turning an Islamic eco-theology into praxis (Amri, 2014). Both organisations offer their institutional backing to local communities seeking to combat deforestation, advance more climate-resilient livelihoods – especially farming – and transition to more efficient use of energy.

Chapter summary

Religious faiths offer illuminative accounts of the human condition. Since the idea of climate change is fully embedded within human culture and the experience of being human, it is inevitable that the religious mind engages with the idea of climate change. The full range of meanings of climate change will not fully be exposed without careful consideration of the many varieties of religious engagement with the idea and how they vary from place to place. Religious cosmologies, doctrines, myths, experiences, and subjectivities shape the nature and credibility of different knowledge claims about climate: What is happening to climate and why. These aspects of religious life also exert a powerful influence in guiding individual and communal ethical and social behaviours in response to climate change. Amitav Ghosh (2016) recognises the animating power and moral energy that can emerge from religious narratives and movements. Unlike many secular thinkers, Ghosh argues that without this animating power of religion – and its sense of the sacred – there is little hope for developing any adequate response to the challenges of climate change.

Given their kaleidoscopic nature, these religiously shaped encounters with climate change are things about which geographers of religion have much to say. This would include the following. Different religions offer more or less formal confessional statements about what is at stake with climate change. These statements provide guidance for the faithful about what climate change signifies according to specific faith traditions and ethical duties. Public discourse and policy debates frequently involve religious actors or draw upon religious imagery and narratives in the search for cultural authority. Such religious interventions in public debates commonly place the challenges of climate change within larger narratives of human dignity, duty, purpose, and even cosmic destiny. Conditioned as they are by local and national cultures, religious identities correlate in complex ways with political attitudes to climate change.

***

This chapter ends Section 2 of the book. Chapters 3, 4, and 5 examined how different combinations of political ideologies and normative commitments pass contrasting judgements on the scientific facts of climate change. From these judgements are developed three different accounts of what climate change means for political action in the world. Chapters 6, 7, and 8 then highlighted three "more-than-science" ways of imputing meaning to the idea of climate change, pursued respectively by subalterns, creative artists, and the world's religious faiths. These latter three chapters show how "science-based" accounts of climate change can, respectively, be supplanted, reimagined, and transcended.

Further Reading

Perhaps the most far-reaching interpretation of climate change from any single faith position is that offered by **Pope Francis** in his *Laudato Si'. On Care for Our Common Home* (Melville House Publishing, 2015). This encyclical is well worth reading. **Katherine Wilkinson's** *Between God and Green: How Evangelicals are Cultivating a Middle Ground on Climate Change* (Oxford University Press, 2012) explores what climate change signifies to a different Christian tradition, American evangelicalism – a theme also explored in **Robin Veldman's** *The Gospel of Climate Scepticism: Why Evangelical Christians Oppose Action on Climate Change* (University of California Press, 2019). Also from an American religious perspective is **Peter Thuesen's** *Tornado God: American Religion and Violent Weather* (Oxford University Press, 2020), which offers an excellent historical account of how extreme weather is interpreted by religious culture. Beyond Christianity is **Fazlun Khalid's** *Signs on the Earth: Islam, Modernity and the Climate Crisis* (Kube Publishing, 2019) and **Robin Veldman, Andrew Szasz**, and **Randolph Haluza-Delay's** edited collection of essays *How the World's Religions are Responding to Climate Change: Social Scientific Investigations* (Routledge, 2014). This latter collection shows how different faith traditions in different geographical settings engage with the idea of climate change. **Evan Berry's** edited book *Climate Politics and the Power of Religion* (Indiana University Press, 2022) brings together some valuable insights at the intersection of religion and climate politics, also with case studies from around the world. Two books that approach climate change from a distinctly theological perspective are **Michael Northcott's** polemical *A Political Theology of Climate Change* (SPCK, 2014) and a collection of essays from more diverse theological standpoints in **Michael Northcott** and **Peter Scott's** *Systematic Theology and Climate Change: Ecumenical Perspectives* (Routledge, 2014). An idiosyncratic but illuminating personal view of the relationship between climate change science and religion can be found in some of the chapters of **Sir John Houghton's** autobiography *In the Eye of the Storm: The Autobiography of Sir John Houghton* (Lion, 2013).

QUESTIONS FOR CLASS DISCUSSION OR ASSESSMENT

Q8.1: To what extent do you agree that religions can offer societies what Jürgen Habermas calls "the missing moral commitments" (Habermas et al., 2010) for dealing effectively with climate change?
Q8.2: Why do you think the United Nations' IPCC assessment reports do not include a section or chapter on "religion and climate change"?
Q8.3: What reasons might account for more than twice the proportion of Hispanic Catholics in the USA being "very concerned" about climate change compared to White Catholics (see Figure 8.2)?
Q8.4: Organisations with "religious affiliations possess the ability to mobilize people in far greater numbers than any others" (Ghosh, 2016: 160). Why might Ghosh place such faith in the mobilising power of religion? Do you agree with him?
Q8.5: Do religious leaders have more or less responsibility compared to political leaders for speaking out publicly about climate change?

Section 3

CLIMATE CHANGE TO COME

In Section 1, I explored the historical and cultural origins of the idea of climate change and how, over the past two centuries, this idea became the subject of scientific investigation. Section 2 then provided a series of investigations into the ways in which, over the past 30 years, different political and cultural meanings have been imputed to the idea of climate change. The six chapters in that section each described a broad interpretative stance on how and why climate change matters for different people in different places. The first three positions I described as "science-based" and the second three approaches as "more-than-science". Now, in this final section of the book, I conclude with two chapters that examine the complex relationship that societies and people have with the climatic future, through the respective lenses of governing and futuring climate.

In Chapter 9: *Governing Climate*, I describe the multiscale architecture of climate governance that is developing in response to the idea of climate change, an architecture some political climate theorists call "a regime complex". This complex embraces a wide variety of state, nonstate, and trans-state institutions and actors: Emissions markets, investment panels, courts and tribunals, company boardrooms, trade associations, town councils, design codes, social movements, and financial regulators. Climate governance has become much more than governing "the climate", narrowly defined. Rather than denoting a specific new set of political institutions or processes, governing climate has become a synecdoche for governing the future. As well as turning climate into an object of governance, the idea of climate change carries human hopes, anxieties, and fears for the future. Climate change has the ability to create both dystopic and utopic imaginary future worlds.

In Chapter 10: *Climate Imaginaries*, I show how different imaginaries are realised through different futuring practices: Scenario planning and scientific modelling and metaphors and speculative fiction. The distinctions between utopic and dystopic climate futures – and the political struggles to realise them – are nowhere better illustrated than through the imaginary of geoengineered technoclimates. Climate change will forever be in the human imagination, as much as it will continue to reshape physical and social worlds.

The "reformers", "contrarians", and "radicals" of Chapters 3, 4, and 5 will approach questions of climate governance in different ways. These three broad groups have (very) different objectives with respect to climate change and will pursue these objectives through different means. Likewise, the subalterns, creative artists, and religious communities of Chapters 6, 7, and 8 will interrogate the objectives, institutions, and power dynamics of climate governance differently, bringing different styles of protest, challenge, and support to bear on these various processes. It is also the case that the different positionalities on climate change examined in Section 2, will inspire and respond to different climate futures brought into being by the range of futuring practices explored in Chapter 10. The idea of climate change has a fecundity that far exceeds the ability of any one of these positions on climate change to master and control.

9

GOVERNING CLIMATE

. . . Climate change governed . . .

Introduction

At the 70th Plenary Meeting of the 43rd Session of the United Nations, held on 6 December 1988 in New York, the world's governments passed Resolution UNGA 43/53. This resolution called for "The conservation of climate as part of the heritage of mankind" and was submitted by the Government of Malta, which during the Cold War was a signifier of the non-aligned world. UNGA 43/53 required governments and intergovernmental organisations to "make every effort to *prevent detrimental effects on climate* and activities which affect the ecological balance" (emphasis added) and called upon nongovernmental organisations and industries to also play their part. The resolution paved the way, three and a half years later, for the signing of the UN's Framework Convention on Climate Change (UNFCCC, or "the Convention") at the Earth Summit in Rio de Janeiro. It was significant of course that this same resolution mandated the newly formed United Nations' Intergovernmental Panel on Climate Change (IPCC) to assess the state of knowledge about the science and impacts of climate change (see Chapter 2).

Resolution UNGA 43/53m – and the UN's decision to work towards a climate change convention – may be seen as a key moment in the history of the idea of climate change. Climate change had long been an imaginative idea that interpreted the human experience of the world (Chapter 1) and, more recently, it had become subject to the forensic investigation of scientists seeking to understand and explain the physical processes of planetary change (Chapter 2). But climate change was now also to be regarded as a pathological condition of modernity that threatened "the heritage of mankind". The resolution brought by the Maltese Government in 1988 to the world's nations gathered in New York made clear that steering the world away from the dangers of unbridled climate change would require a mobilisation of the whole of humanity and its governmental capabilities, both state and nonstate. For the first time, climate – more specifically, *global* climate – was to be the subject of explicit and purposeful intergovernmental intervention.

The ambition and structures of global climate governance have therefore been around for about 30 years. The formal intergovernmental institutions that Resolution UNGA 43/53 eventually brought into being – the IPCC, the Convention, the Conferences of the Parties to the Convention (and their associated paraphernalia) – can easily become the focus of attention when thinking about how climate governance works and when evaluating how successful it has been. These institutions are certainly important and must not be ignored. But as we shall see in this chapter, the agents of climate governance extend today well beyond these formal institutions. These agents include, among others, building inspectors, venture capitalists, media producers, trades unionists, monks, aviation authorities, professional sports clubs, farming extension officers, public celebrities, and national energy regulators.

The reasons for this proliferation of actors in the governing of climate are twofold. First, is the fact that nearly every sphere of human activity alters, to some degree, the physical properties of the atmosphere. Every human activity can affect the climate. But, second, this proliferation also reflects the divergent cultural discourses and political conflicts surrounding the idea of climate change that have been explored in previous chapters. Climate change means different things in different social worlds. It inspires many different personal reactions and political projects. It would therefore be surprising if such conflicting interests and diverse aspirations could be resolved through a centralised and carefully orchestrated intergovernmental process. Securing "the conservation of climate" for the benefit of humanity is no easy matter. As argued in the Prologue, climate change is a wicked problem. The idea of climate change is too fractal, too multifarious – it has unleashed too much political conflict in the world – for such an outcome ever to be possible. It is a modern conceit to think that the world's governments could satisfactorily regulate global climate. And yet this was the ambition set in motion in 1988 by the Government of Malta.

Given this plurality of engagements with climate change governance, it is not surprising that attendance has swollen at the annual Conferences of the Parties (COPs) to the Convention – sometimes referred to as the "international climate negotiations". The first COP was held during a ten-day period in the early spring of 1995 in the recently reunified city of Berlin. This venue was symbolic of the new spirit of internationalism that prevailed in the first post-Cold War decade. COP-1 in Berlin attracted governmental delegations from 117 parties (nation states) to the convention, plus 53 observer states. Participants numbered in the few thousands. In the subsequent 25 years, the annual COPs have become increasingly cosmopolitan gatherings. Governmental officials, UN diplomats, and national politicians are now in the small minority. The tens of thousands of attendees – in 2015, COP-21 in Paris attracted around 45,000 participants, policed by 30,000 security staff – now comprise lobbyists, activists, priests, Indigenous campaigners, sceptics, artists, business leaders, scientists, social science researchers, journalists, film commissioners, and many more.

And it is not just who is there that matters. It is what takes place. These annual gatherings of the "climate concerned" – the sequence broken only in 2020 because of the SARS-CoV-2 pandemic – enable much more than arcane diplomatic deliberations over Article 2.1(c), sub-clause 2, or some such concern. Yes, such micro-scrutiny of international negotiating texts occurs. But these cosmopolitan gatherings also enable political lobbying, civic protests, media briefings, scientific announcements, film premieres, technology exhibitions, artistic performances, and art exhibitions, for example ArtCOP21 in Paris as we saw in Chapter 7. COPs have become as much a world festival of science, culture, and human diversity and vitality as they are formal intergovernmental climate negotiations.

The political scientist Carl Death describes such summits as "theatrical dramaturgy". The symbolic aspects of summitry he contends, "are not merely sideshows to the main business of negotiations, but are rather essential to the manner in which [such] summits [now] govern" the conduct of global climate politics (Death, 2011: 6). Stefan Aykut also adopts this Foucauldian perspective on **governmentality** but pushes it further. "Instead of opposing 'symbolic' politics to a hypothetical 'real' politics, we should accept that symbols and narratives form part and parcel of contemporary liberal governmentality" (Aykut et al., 2020: 14). In this new world of theatrical governance, Aykut claims, "performances, symbols and narratives appear to be just as important as the production of rules, institutions and instruments" (p. 3).

***

This chapter analyses the complex architecture of global climate governance brought into being by the multifaceted idea of climate change that the previous chapters have explored. The first two pillars of the climate regime were designed around state actors and interests: The 1992 Framework Convention and the subsequent 1997 Kyoto Protocol. And although the formal outcomes of subsequent COPs are the result of inter*governmental* negotiations – i.e., negotiations by state representatives – global climate governance has evolved into what some political climate theorists call "a regime complex" (Keohane & Victor, 2011). This complex embraces a wide variety of state, nonstate, and transstate institutions and actors. Climate governance is enacted through emissions markets, investment panels, courts and tribunals, company boardrooms, trade associations, town councils, design codes, social movements, and financial regulators as much as it is advanced by state governments. And climate governance is about much more than governing "the climate", narrowly defined.

Rather than denoting a specific set of institutions or processes, global climate governance has become a synecdoche for governing the future, in effect governing the future of human life on a finite Earth.

To offer insight into this "regime complex", the chapter is structured around three central questions. First, what exactly is the *object* of climate governance? Rather than governing climate itself, we shall see that governance comprises the regulation of technologies and human behaviours (climate mitigation) and enhancing the resilience of societies (climate adaptation). Second, which *actors and agents* are deemed to have legitimacy and capacity to govern climate? The answer to this question reveals an accelerating transition from a state-centric to a polycentric system of climate governance. Third, by what *modes* of intervention is future climate governed? This question is answered by looking briefly at five mechanisms of governance: Standards and certificates, carbon markets, citizens' assemblies, courts, and climate services. The answers to these questions are shaped by the different meanings attached to climate change by the different political, social, and cultural actors outlined in earlier chapters. They each have different ends in sight and preferred means to achieve them. The chapter draws mostly on the disciplines of *political geography, international relations*, and *science and technology studies.*

What is being governed?

According to the 2015 Paris Agreement, the object of climate governance is to regulate global temperature. "Success" in this governance project would be to limit the increase in global temperature to no more than 2°C above its pre-industrial condition, with an aspiration for this increase to be closer to 1.5°C (see Box 9.1). The world is presently around 1°C warmer than the late nineteenth century. But global temperature is not a physical property of the planet over which governments – or anyone else – have direct control. In treaties like the Paris Agreement and in popular discourse, global temperature acts as a proxy for "other things".

These "other things" might be those phenomena that in some way are physically linked to global temperature – i.e., dangerous meteorological and environmental hazards such as cyclones, heatwaves, ice-storms, torrential downpours, or a rising ocean. It would certainly be desirable to minimise the occurrence or severity of these hazards. But clearly, as with global temperature, the occurrence of, say, damaging ice-storms cannot be directly regulated by purposeful human actions. The chain of causation has too many links and is too uncertain. So, more helpfully for the purposes of governance, these "other things" are activities over which human institutions do have more direct control, for example, energy technologies, land use, human mobility, consumption, and so on. Global temperature is only a useful target of climate governance in so far as it is a credible proxy for the complex relationship between human activities and changing weather. It is worth elaborating these distinctions a little further in the subsections to follow, which focus, respectively, on climate, technologies, human behaviours, and social adaptation as objects of climate governance.

Climate

Global climate is not something over which governments have direct control. This is why the 1992 Convention adopted as its ultimate objective the rather evasive formulation, "avoiding dangerous climate change". In the 30 years since the convention was signed in Rio, this objective has been re-expressed in different ways (McLaren & Markusson, 2020). Limiting the increase in global temperature has been the most salient of these reformulations (see Box 9.1). But other indicators for successful climate governance have been proposed, for example ocean heat content, carbon dioxide emissions, atmospheric carbon dioxide concentrations, global sea level rise (Victor & Kennel, 2014). These are all *climate system* indicators, proposed as measurable targets that reveal the effectiveness – or otherwise – of climate governance.

It is interesting to note what is *not* captured by such indicators. Successful climate governance is not, it would seem, to be measured against human-centred indicators, such as reducing the number of people either dying or left without livelihoods because of climatic hazards or else sustaining the health of ecosystems. While such outcomes may be understood tacitly as secondary benefits of governing climate, such indicators have never been offered as the headline targets of climate governance. This is in contrast to the UN's 17 Sustainable Development Goals (SDGs), where securing such welfare benefits *is* the explicit purpose and goal of policy interventions. The climate governance regime therefore operates primarily as a mitigation paradigm – limiting change to a physical system. The SDGs primarily reflect a welfare paradigm – promoting positive human welfare. This difference in focus is significant and can be traced back to the dominant role played by Earth System science in the 1980s and 1990s in framing the nature of the climate problem and its imagined alleviation.

BOX 9.1 GOVERNING CLIMATE BY GLOBAL TEMPERATURE

The formal adoption of a global temperature target to guide international climate policy development dates back to the mid-1990s. The origins of the "two degrees" target – limiting the rise in globally averaged temperature to no more than 2°C above a late nineteenth-century

(*Continued*)

(1850–1900) baseline – can be traced back to early economic analysis of climate change in the mid-1970s and to ecological analysis of dynamic forest boundaries in the early 1980s (Randalls, 2010). Two degrees was first adopted as a political target in 1996, by the EU Council, before gaining intergovernmental prominence in the Cancun Agreements, which emerged from COP-16 held in Mexico in 2010.

The power of global temperature for animating climate politics – and also climate science – was shown during the negotiations leading to the Paris Agreement in 2015. Two degrees was unanimously agreed in Paris by the world's governments as the formal goal of the agreement but with a further number inserted. It was agreed that governments would "pursue efforts" towards limiting the rise to as close to 1.5°C as possible. But for this number to acquire the necessary epistemic authority to act as an object of governance, 1.5° needed affirmation by scientific analysis. The signatories to the Convention therefore "invited" the IPCC to produce a report to "discover" the benefits of limiting the rise to 1.5° rather than 2°C. As Livingstone and Rummukainen (2020) reveal, the scientific community were "surprised" by this request but nevertheless duly obliged, it being a case of "the research process awkwardly aligning itself with policy demands" (p. 15). The IPCC report *Global Warming of 1.5°C* was duly published in 2018 (IPCC, 2018).

This index of climatic performance – collapsing the complexity and diversity of weather around the world into one (or two) number(s) – has gained powerful political and iconic cultural status. Global temperature *appears* to make climate governable. The power of the target resides in its apparent simplicity. Two degrees of warming offers a precise number to aim for that can be monitored scientifically.[1] Global temperature has been remarkably effective as an object around which scientific research, public policy, and civic activism has each drawn inspiration. It functions as a boundary object (see Chapter 2) in science-policy analysis and in public communication and advocacy.

The "success" of global temperature as an object of governance can be explained by understanding that numbers embody epistemic authority. Modern forms of governmentality require the objects of governance be enumerated (Scott, 1998). Indexing "the health of the climate" to global temperature fulfils this need. It offers a numerical index against which progress towards its achievement can be monitored, seemingly objectively. Global temperature offers a fixed benchmark against which governmental actions and human decisions can be held accountable. It appears as a "natural kind" of knowledge about the world, beyond manipulation by malfeasant political interests. The irony is that the very success of global temperature as the ostensible object of climate governance might well lead to its demise (Lawrence & Schäfer, 2019). The likelihood of limiting global temperature rise to 2°C – let alone 1.5°C – is rapidly diminishing, suggesting that before long "two degrees" will no longer have traction as a credible indicator of successful climate governance.

Read:

Randalls, S. (2010) History of the 2°C climate target. *WIREs Climate Change*. 1(4): 598–605; Livingstone, J.E. and Rummukainen, M. (2020) Taking science by surprise: the knowledge politics of the IPCC Special Report on 1.5°C. *Environmental Science & Policy*. 112: 10–16.

Technologies

In scientific terms, global temperature is a function of the balance of heat entering and leaving the planetary atmosphere and oceans. At its simplest then, global temperature can be regulated in two ways: Either by reducing the amount of heat entering the planet or by increasing the amount leaving. Technologies are available to accomplish both these objectives. One object of climate governance therefore becomes technology. For example, solar geoengineering technologies have been proposed to reduce the amount of energy *entering* the top of the atmosphere. These technologies – such as spraying aerosols in the stratosphere – seek to regulate directly the flow of solar radiation into the planet. They remain presumptive technologies; they do not yet exist. But their mere possibility raises vital questions about who mandates their possible development and, in the future, how their deployment at scale across the planet might be reliably and justly governed (Hulme, 2014b).

To increase the amount of heat *leaving* the planet requires a reduction in the concentration of heat-trapping greenhouse gases in the atmosphere. And since burning fossil fuels for energy is the primary reason for the historical accumulation of greenhouse gases, debates about energy technology have long been central to the politics of climate change. Regulate the technologies by which the modern world acquires its energy and one can control fossil fuel emissions. Eliminate such emissions and one can go a long way to arresting climate change. Energy technologies also therefore become an object of climate governance. Such regulation may be achieved in numerous ways, some of which were discussed in Chapters 3, 4, and 5: Setting decarbonisation targets for national jurisdictions or individual companies to pursue; public investment in low carbon energy; implementing policies to stimulate private innovation in clean energy technologies; taxing carbon emissions or fossil fuel extraction; withdrawing from fossil fuel industries the social licence to operate. Reformers, contrarians, and radicals position very differently on these various options.

A related object of climate governance is technologies that seek to remove carbon dioxide directly from the atmosphere – carbon dioxide removal technologies. We examined some of these in Chapter 3 in the context of reformed modernism (see Table 3.2). The governance challenges of CDR technologies are not just technical, economic, and social – can they be designed, are they competitive, and can they be deployed without generating social conflict? The challenges have as much to do with the way in which their promise to extract or absorb carbon dioxide from the atmosphere intersects with the difficult politics of conventional climate mitigation (Figure 9.1). CDR technologies offer a powerful socio-technical imaginary (see Chapter 10) in which ambitious net-zero emissions targets can seemingly be met without the need to eradicate fossil carbon from the energy system. Most of these technologies do not yet exist at scale, even though their imagined effects can be simulated within models of the climate-energy-society system. These technologies therefore exert a powerful effect on the governance of climate even though, according to some analysts, their

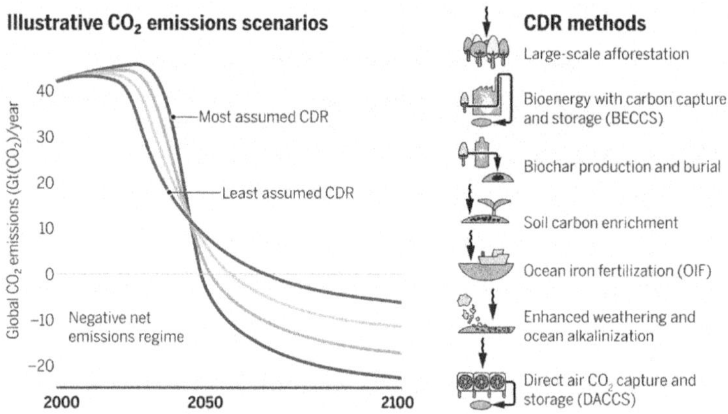

Figure 9.1 How hypothetical CDR technologies shape climate scenarios. Many of these CDR methods remain untested at the required scales.
(Source: Lawrence & Schäfer, 2019; reproduced with permission).

promise to reduce future atmospheric concentrations of carbon dioxide is somewhat akin to "magical thinking" (Geden, 2015).

These three cases of governing climate by governing technology raise the question of responsible innovation. This question was originally posed in the 1980s by technology analyst David Collingridge (1980). His "technology control dilemma" consisted of two observations. First, the social, economic, and environmental impacts of a technology – and hence its desirability – cannot easily be predicted until the technology is extensively developed and widely used. But second – and here is the dilemma – once a technology *becomes* developed and widely used it becomes difficult to change or control. Collingridge's dilemma applies to many of the technologies mentioned previously in relation to climate governance. Today, these concerns are more likely expressed using the language of responsible (research and) innovation (see Genus & Stirling, 2018). At the heart of responsible innovation lies the question of which people and whose interests are to shape the processes and scrutiny of technology development and deployment. Collingridge's "technology control dilemma" applies equally, among many other technologies, to solar geoengineering, hydrogen fuels, nuclear power, driverless cars, direct air capture, and carbon capture and storage.

Human behaviours

People and their behaviours also comprise an object of climate governance, especially people's preferences around consumption, mobility, and procreation. Such a focus attends to the possibilities for state or other agencies to intervene in social and cultural life in ways that influence and steer individual choices in directions deemed to be less damaging for the climate. Climate thus becomes governed to the extent that people's behaviours are governed. This governance ambition raises numerous political and ethical questions about the relationship between state and individual, freedom of choice, and definitions of wealth and well-being. These are of course enduring questions for societies, which social and political philosophers have long grappled with, but the idea of climate change animates them in new and challenging ways. With respect to planned interventions that seek to influence human behaviours, these questions are explored in Richard Thaler and Cass Sunstein's book *Nudge: Improving Questions about Health, Wealth and Happiness* (2009). Again, reformers, contrarians, and radicals will position very differently on these political and ethical questions, while subalterns, artists, and religious communities will inflect these arguments in distinctive ways.

Klintman and Boström (2015) offer a helpful classification to understand the different ways in which "citizen-consumers" become active in climate governance (Table 9.1). Their classification makes two distinctions. The first

Table 9.1 Types and examples of actor- and structure-oriented climate mitigation strategies.

	Focus on citizens' purposes	*Focus on structural arrangements*
Climate mitigation as explicit, primary goal	Buying carbon-labelled goods	Constructing carbon labels
	Carbon-sensitive diets	Carbon tax arrangements
	Switching to green energy supplier	Carbon rationing
		Urban mobility policies
Non-climate primary goals	Buying organic and other eco-labels	Mobility management
	Vegetarianism, veganism	Urban planning
	Health bikers	Policies for "green jobs"
	"Quality of life" movements	Public procurement policies
	Economics of sharing	

(*Source*: Adapted from Klintman & Boström, 2015).

is between behaviours regulated with climate mitigation – the reduction of greenhouse gas emissions – as an explicit goal, rather than those with mitigation as a side-effect of securing other policy goals. The second distinction is between the purposeful actions of *individuals* versus the institutional and structural arrangements of *a society* that direct or constrain those individual actions. This fourfold classification helps isolate the different types of policy interventions that might alter human behaviours in ways that may be deemed "climate-friendly". It suggests different ways of governing behaviour in order to govern climate. For example, interventions such as carbon-labelling of consumer products are focused on individual citizens' choices with an explicit focus on climate mitigation. In contrast, rethinking the purposes and outcomes of urban planning alters the spatial arrangements of cities in which people live, move, and make their choices. The concerns of urban planners encompass not just energy use and carbon emissions but broader social objectives such as mobility, justice, and well-being.

One human behaviour that rarely emerges as a direct object of climate governance is the decision whether or not to procreate. To what extent should the decision to have children be subject to state influence or be mediated through other social collectives or cultural narratives of responsible climate care? Is procreative autonomy an inviolable individual freedom beyond the reach of climate governance? Examples of state-led population policies are of course well-known – coercive sterilisation programmes in India in the 1970s and China's one-child policy in the 1980s. But these policies were not undertaken with climate change in mind. And even though attention more recently, mostly in Western nations, has been drawn to "responsible" procreative choices specifically in the context of climate change, this is rarely framed as a matter of climate governance.[2]

Adaptation

The previous objects of climate governance either shield the planet from additional heating or else involve reducing the impact of diverse human activities on the atmosphere. These are strategies of, respectively, climate remediation and mitigation. A different focus of climate governance that has gained in prominence in the twenty-first century concerns the adaptive capacity and resilience of societies. In other words, how may societies, individuals, and ecosystems be "steered" so as to make them less susceptible to *harms* from future climatic hazards? Or, conversely, how might they be enabled to *benefit* from new opportunities offered by a changing climate? Adapting to weather and climate variability and extremes has been an innate process of adjustment, innovation, and experimentation over countless human generations. People in different cultures have found numerous ways of taming the dangers of climatic extremes, of living safely and purposefully with their weather (see Chapter 1). Some of these have proven more successful than others. One might therefore wonder whether the idea of climate change in any way requires new modes and processes of "governance" in order to secure what has been historically a largely spontaneous activity.

The answer is becoming increasingly obvious. Given the realities of a changing climate, adaptation processes that are more proactive and precautionary *are* deemed desirable by a widening array of individuals and institutions. Rather than *reacting* to weather and climate hazards, climate change adaptation is based on *anticipatory* actions – adjustments made to societies ahead of realised changes in climatic conditions. The question then becomes who takes responsibility for such interventions: The state, other collective agencies including commercial entities, or the individual? These responsibilities are distributed differently within and between nations and constitute what Benjamin Sovacool and Björn-Ola Linnér describe in their eponymous book (2016) as the "political economy of climate change adaptation". The precise *goal* of good adaptation governance is also disputed. Is it to reduce loss of life and economic damage, to enhance sustainable development, or to build more adaptive and resilient societies? Reformers, contrarians, and radicals will have different answers to this question. Sovacool and Linnér show that purposeful climate adaptation projects can produce unintended and undesirable outcomes, reflecting the different political regimes, institutional capacities, and cultural norms of different countries.

Who is governing?

Beyond the question of *what* is being governed, it is necessary to also consider *who* is governing. The significance of Resolution UNGA 43/53 in 1988 was that it set in motion a chain of events leading explicitly to nation-states taking on some shared responsibility for governing the climate, albeit in some cases reluctantly. Dealing with the hazards of climate and the risks of climate change was no longer to be left to individuals or private collectives, such as communities, NGOs, or businesses. As with homeland security, the strength of the economy, or public health, climate change was henceforth to be the business of state governance.

State-centric global governance

The founding assumption of the 1992 Framework Convention was that states would negotiate, agree, and then execute their responsibilities for governing the climate. This is not to say that all states had *equal* responsibility. Far from it. The history of climate governance is deeply rooted in the geopolitics of power and territoriality. There is an irony here. On the one hand, the Earth System science which underpins the convention portrays the climate of the world as a unified system. And yet, on the other, various treaties negotiated between 197 nation-states under the convention have repeatedly divided the world into different blocs. Initially, this was an echo of the Brandt Commission's division of the world in 1980 between "the Rich North" and "the Poor South", the so-called Brandt line (Figure 9.2a). This division was enshrined in the Convention (in Article 3) through the principle of "common but differentiated responsibilities and respective capabilities". Thus the 1997 Kyoto Protocol differentiated between so-called Annex-1 (developed nations) and non-Annex-1 countries (Figure 9.2b). Most members of the former group took on specific emissions reduction targets; the members of the latter did not.

The shift to the design of the 2015 Paris Agreement has been to weaken this principle – or at least to accommodate a much looser interpretation of it. In the agreement, *all* nations have agreed to define their own **Nationally Determined Contributions** (NDCs) to the collective effort to reduce emissions (Figure 9.2c). As of February 2021, only six of the 197 Parties to the Convention had not ratified the PACC,[3] the remaining 190 having agreed to designate and pursue their own NDCs. Beyond the different groupings enshrined in these headline treaties, numerous other country groups or "climate clubs" have been influential at various times within the negotiating process. This

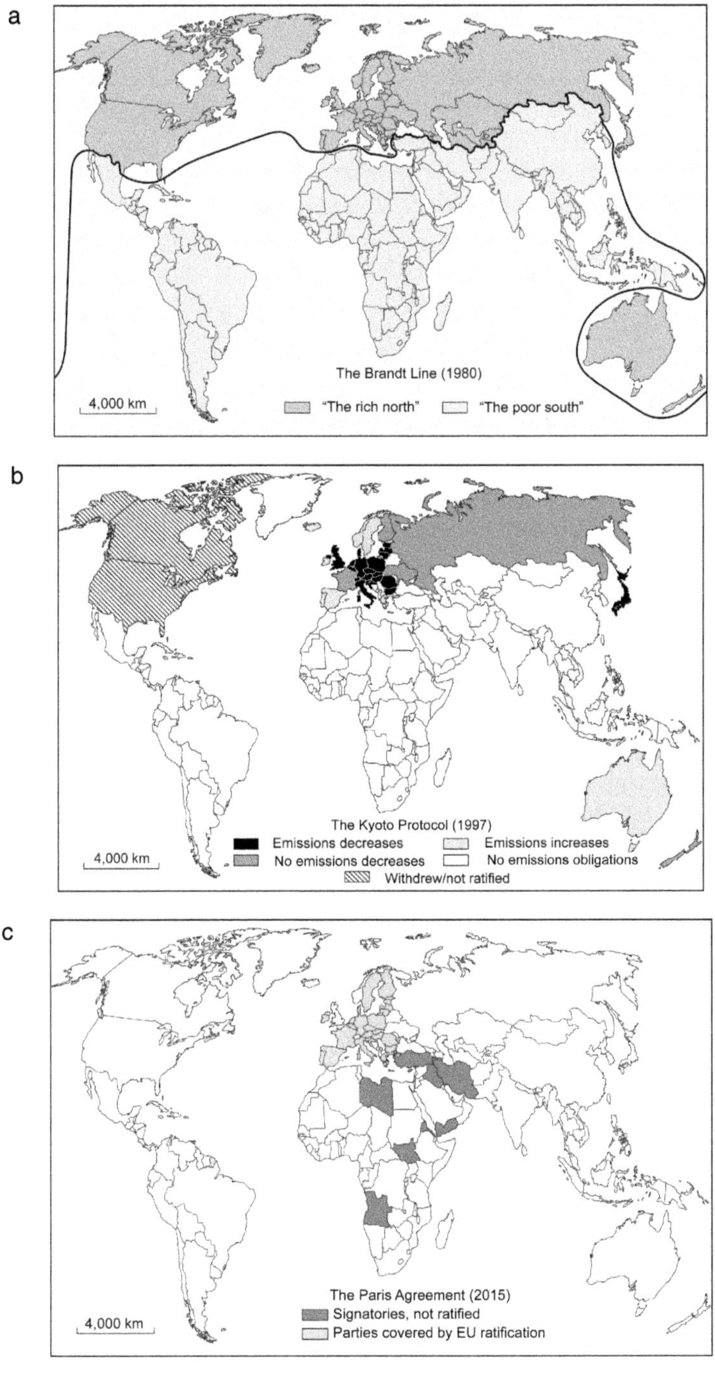

Figure 9.2 The spatial politics of climate governance: a) the binary world of the Brandt Commission; b) countries' emissions obligations under the Kyoto Protocol; c) countries that have not yet ratified the Paris Agreement.

includes groups such as the G77 + China, AOSIS, the BASIC group, the Arab Group, the Coalition for Rainforest Nations, OPEC, the Environmental Integrity Group, and so on (Gupta, 2014).

The 30 years of climate governance set in motion by the UN's 1988 resolution has therefore mapped and remapped the world in different ways. Nevertheless, in terms of legal treaties it has been a story of state-centred governance. It is states alone that have the legal authority to negotiate and enter into treaties under the UN system. Perhaps the epitome of state-centred global climate governance has been the attempted mobilisation of the UN's Security Council (UNSC) in support of this goal. Climate change and its potential implications for international security was first introduced to the UNSC in April 2007 under the UK presidency. The debate was framed by the UK Chair as being about "our collective security in a fragile and increasingly interdependent world". Such cosmopolitan high-mindedness was followed just over two years later by the acrimonious failure of the Copenhagen negotiations at COP-15. Since then, there have been six further debates in the UNSC about climate change, proposed, respectively, by Germany (in 2011), New Zealand (2015), Senegal (2016), Sweden (2018), the Dominican Republic (2019), and the United Kingdom (2021). Little of substance has emerged from these UNSC interventions.

Even less has been realised from the ambitious vision for Earth System Governance. This vision takes inspiration from the collective achievements of Earth System science and proposes an integrated, multi-scale system of governance by which all human-induced transformations of Earth System are regulated. The German political scientist Frank Biermann has laid out perhaps the grandest version of such a unified and orchestrated design of global climate governance in *Earth System Governance: World Politics in the Anthropocene* (2014). This cosmopolitan vision is to be delivered through a revitalized United Nations, one that would include a new World Environment Organization and a new UN Sustainable Development Council. Biermann's state-centred – but tightly orchestrated – system of climate governance proposes innovative systems of qualified majority voting in multilateral negotiations. It also allows for representation in global decision making from civil society and from scientists. This vision for governing the world's climate seems much further away from being realised than when it was published in 2014.

Polycentric climate governance

A very different way of seeking to understand climate governance decentres the role of the state. Rather than coordinating state actions through multilateralism – in other words *monocentric governance* – this alternative line of thinking recognises that multiple semi-autonomous authorities are active at many different scales. These nonstate authorities operate above, below, and beyond the state. They include businesses, municipalities, trade associations, the courts, civic movements, religious bodies, banks, investors, and emissions markets, to name a few. This thinking resonates with the theory of *polycentric governance*, an idea most closely associated with the Nobel Prize–winning economist Elinor Ostrom (1933–2012). Ostrom developed her theory of polycentricism through work she conducted in the 1970s and 1980s on common-pool resources – such as agriculture and fishing – in many countries around the world. Ostrom's argument is best developed in *Governing the Commons: The Evolution of Institutions for Collective Action* (1990). She advances the claim that polycentric governance systems offer institutions greater scope for policy experimentation than could be secured under state-centred regimes. Her theory suggests that in a polycentric governance system, a greater range of benefits accrue for multiple political actors across different scales than would be the case in a monocentric system.

Polycentricity is a useful analytical approach for describing and understanding how climate governance has developed over the years. And it resonates with other descriptions that political scientists have adopted, such as transnational climate governance, networked private and hybrid governance, and regime complex theory (Figure 9.3). For example, the idea of transnational climate governance is defined as "governance at multiple scales by a diversity of state, sub-state and non-state actors" (Dingwerth & Green, 2015: 153), which is a good enough description of polycentricity. These various descriptions of climate governance capture the move away from state negotiation and

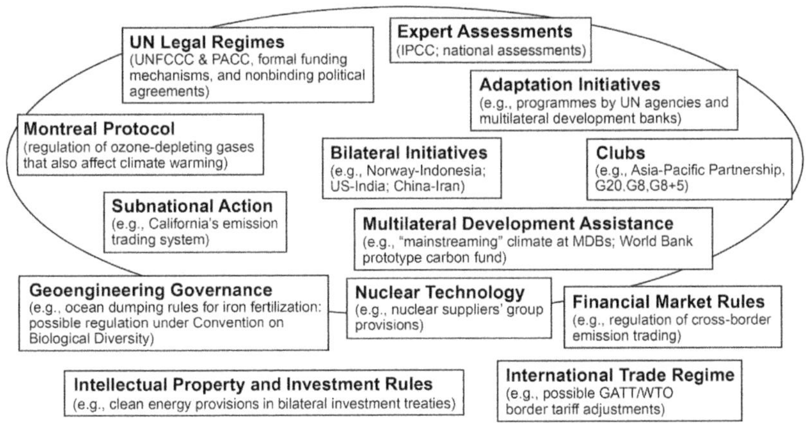

Figure 9.3 The regime complex for managing climate change.
(*Source*: Redrawn from Keohane & Victor, 2011).

coordinated actions under the UNFCCC in favour of a diverse, looser, and less hierarchically orchestrated set of governance actors.

Polycentric governance of climate would appear to be a more suitable description of what has emerged over the past 10 to 15 years than the grand ambitions and rhetoric of Earth System Governance. Early efforts made under the convention were to secure coordinated and sanctioned multilateral state-centred treaties. These efforts largely failed, culminating in the regressive outcome of COP-15 at Copenhagen in 2009. Climate governance has instead become complex, fragmented, and decentralized, operating with very limited central coordination. The Paris Agreement allows climate governance to unfold from the "periphery", from the "bottom", rather than be executed from above and through the state. National governments are asked only to specify their (voluntary) NDCs, which countries are free to design in ways that are consonant with their own institutional, economic, political, and moral characteristics and interests. There is no enforcement regime, merely so-called pledge-and-review and global stocktakes every five years.

The agreement therefore explicitly recognises the many loci of governance that exist beyond the nation state. It embodies a very different conception of power and governance than did, for example, the Kyoto Protocol. Its headline ambition is to limit global warming to 2°C. Yet the architecture of the agreement tacitly concedes that this outcome cannot be centrally secured. Neither state- nor UN-centric manifestations of power – nor indeed the workings of the market, nor the surprises wrought by cultural, economic, and technological shifts – will ever allow future climate a predictable "safe-landing". As some commentators observe, it is good for democracy to move away from "fantasies about centralized, detached steering leading to sweeping global transformations" (Lawrence & Schäfer, 2019: 830).

This move from monocentric to polycentric climate governance, from state-centred actions to transnational networks of sub-state actors, from the Kyoto Protocol to the Paris Agreement, can be explained in a number of ways (Aykut, 2016). Some see it is as a response to earlier failures in the international negotiations, for example the debacle at COP-15 in Copenhagen in 2009. Others would see it as a growing recognition of the "governance gaps" that cannot be filled by UN-based processes and treaties. Others still would see differently the national self-determination of NDCs and the polycentric governance of climate through overlapping layers of nonstate actors tacitly endorsed by the agreement. They would interpret the former move as a response to a changing world order in which national interests are reasserting themselves against cosmopolitan ones. And they would see in polycentricity a pragmatic response to the complexity of the social world and its deep entanglement with the climate system. It is also a more experimental governance regime, open to accommodating some of the "more-than-science" approaches for understanding climate change that were explored earlier in Chapters 6, 7, and 8.

There are both benefits and risks of polycentric nonstate approaches for delivering climate governance as opposed to more central orchestrated and regulated governance frameworks. Lawrence and Schäfer (2019) argue that a combination of voluntary NDCs and polycentrism means that policies and measures – and attendant social changes – are much more likely to align with local systems of representation and accountability. Similarly, in the context of seeking transformation of carbon-based energy systems, Bernstein and Hoffman (2019) argue that the challenges of decarbonisation are better off imagined through the metaphor of "global fractal", rather than the more traditional "global commons". Others however see significant risks with a polycentric, nonstate climate governance regime. These include the lack of learning between actors, poor alignment between policy outcomes at different scales, lack of accountability, and a lack of attention to questions of justice (Chan et al., 2019). This relationship between climate governance and social justice has been the focus of much of the work of political geographer Chuks Okereke over the past 15 years (see Box 9.2).

BOX 9.2 CHUKWUMERIJE OKEREKE (B.1971) – CLIMATE GOVERNANCE AND JUSTICE

Chukwumerije (Chuks) Okereke is a political geographer and Director of the Centre for Climate Change and Development at the Alex Ekwueme Federal University Ndufu-Alike Ikwo (AEFUNAI), in Ebonyi State, Nigeria. Prior to taking up this position in 2019, he had been Professor in the Department of Geography and Environmental Science at the University of Reading, where he also served as Co-Director of the Climate and Justice Centre. After completing a BSc in industrial chemistry from the University of Nigeria, Nsukka, Okereke secured an MSc in Environmental Management from Saxion University of Applied Sciences in the Netherlands. His PhD was awarded by the University of Keele in 2005 for a thesis titled "Conceptions of Justice in Multilateral Environmental Agreements: A Critique of Dominant Approaches in Relation to Sustainable Development and Regime Analysis". He subsequently held research and academic positions in the UK, including in the Tyndall Centre for Climate Change Research, the Smith School for Enterprise and Environment at Oxford, and a Leverhulme Fellowship on "the governance of low carbon development in Africa".

(Continued)

Okereke's work is devoted to advancing policies, strategies, and institutional arrangements that can address climate change and natural resource degradation in Africa. These concerns are placed in the context of sustainable development and Africa's structural economic transformation. Okereke emphasises connections across scales of governance – from local through national to international dimensions – with a strong emphasis on justice and ethics. He applies this latter emphasis to the analysis of corporate climate strategies and the political economy of climate governance, topics about which he has published widely, including *Global Justice and Neoliberal Environmental Governance* (2008). Okereke has also managed in Africa and other world locations complex projects that sit at the interface of governance, climate mitigation, adaptation, and low carbon development. These include projects for the World Bank and the UN Economic Commission for Africa and, in 2009, the first national low-carbon plan in Africa, Rwanda's Green Growth and Resilience Project.

Okereke's move back to his native Nigeria in 2019 was significant for a number of reasons. He has there established and is directing a new Centre for Climate Change and Development. From this position in Africa, Okereke is able to promote research and advocacy that bridges climate governance, African sustainable development, and global justice, and also help develop a new cadre of African climate governance experts.

Read:

Okereke, C. (2008) *Global Justice and Neoliberal Environmental Governance: Ethics, Sustainable Development and International Cooperation*. Abingdon: Routledge; Okereke, C. and Agupusi, P. (2015) *Homegrown Development in Africa: Reality or Illusion?* Abingdon: Routledge; Okereke, C. and Coventry, P. (2016) Climate justice and the international regime: Before, during, and after Paris. *WIREs Climate Change*. 7(6): 834–851.

How are they governing?

The emergence of transnational climate governance with its multiplicity of actors makes it necessary to think very broadly about the modes by which climate governance is accomplished. Political geographer Harriet Bulkeley has explored this question in *Accomplishing Climate Governance* (2016), using the example of the UK. As Bulkeley explains, the question of accomplishment is important. "It is not only a matter of asking how, why, by and for whom climate change is governed", she observes, "but of interrogating how *what it means to govern* is constituted in relation to climate change" (p. 8, emphasis added). Climate change is a powerful and important idea not only because it presents humanity with a series of challenges to its future well-being. It is also important because it is a powerful idea that stimulates new thinking about how to better organise societies. The idea of climate change is generative of innovation, experimentation, and opportunity as much as it defines a specific problem to be solved. This is an important feature of climate change to which we will return in Chapter 10. To illustrate this generative role of climate change, the subsections that follow highlight five modes of climate governance that have emerged across different scales and jurisdictions: Standards and certification, carbon markets, citizens' assemblies, judicial courts, and climate services.

Standards and certification

One of the earliest – and still common – mechanisms for governing emissions of greenhouse gases was the introduction of environmental standards and certification. Sometimes called "environmental certified management standards", these can operate under either statutory or voluntary regimes. Two of the most common such international standards relevant for climate change are ISO 14001 and the Forest Stewardship Council (FSC) certification. ISO 14001 seeks to standardise good environmental practice across industry and civic organisations, for example with respect to the introduction and management of energy efficiency technologies. It is usually implemented as statutory legislation. Statutory standards are often favoured by command-and-control regulatory regimes in centralised political jurisdictions. China would be a good example. Command-and-control regimes are characterised by centralised state regulation of what is and is not permitted by an industry or other economic activity. In China, for example, the overwhelming majority of regulatory instruments are implemented through legal standards aimed at the prevention and control of pollution (Qin & Zhang, 2020).

FSC certification frequently operates under more voluntary regimes, driven in part by consumer pressure applied to international supply chains. As with ISO 14001, it is intended to ensure implementation of sustainable practices worldwide and the avoidance of unnecessary waste across forest-related industries. A similar example is Fairtrade International's Fairtrade Climate Standard (FCS). This has been created to support smallholders and rural communities to develop in more sustainable ways and to facilitate their participation in market-trading. The FCS Standard issues Fairtrade Carbon Credits for smallholders in developing countries to gain access to carbon markets by providing information and facilitating local training. This certification generates carbon credits that can be sold in the voluntary carbon market to buyers and consumers wishing to make a positive impact on the twin issues of climate change and development. Certification standards have therefore become an important part of the governance of carbon markets.

Carbon markets

As we saw in Chapter 3, pricing carbon can be implemented in two main ways: Through state-sanctioned carbon taxes or through market-based trading systems. Both pathways can be understood as modes of climate governance, but the focus here is on market-based institutions. The idea that carbon markets are an effective policy tool for incentivising reductions in emissions first took root towards the end of the last century. The ideology behind carbon markets is derived from welfare economics and was consistent with the neoliberal economic paradigm gaining international prominence during the 1990s. Carbon markets were written into the 1997 Kyoto Protocol as the central policy tool for climate governance and the first fully functioning large-scale carbon market – the EU's Emissions Trading Scheme (ETS) – commenced in 2005. There are now over 20 regional or national carbon trading systems worldwide. For example, South Korea's ETS launched in 2015 and is now the world's third largest carbon market and covers 70 per cent of its domestic carbon emissions.

The desirability and effectiveness of carbon markets as a mode of climate governance has always been contested. Carbon markets act as a difference marker between those favouring either incremental (Chapter 3) or radical (Chapter 5) change in response to climate change. For those favouring carbon markets, the most compelling argument is that prices matter. By putting a price on carbon and then allowing emissions permits to be traded, it is claimed that markets will respond by innovating and delivering new lower carbon – and cheaper – technologies. Markets are also seen by their advocates as more politically expedient than a carbon tax. The attempts in Europe in the early 1990s to introduce a carbon tax failed. The Obama Administration in the USA tried arguing in the early 2010s that an economy-wide carbon tax was better for the economy than the alternative Clean Power Plan. It too failed. Where carbon tax proposals *have* survived – for example in Scandinavian countries – they are usually weakened through exemptions (Sato & Laing, 2020). The flexibility of market institutions is also in their favour. Carbon markets can plausibly be implemented at many different scales. City-scale schemes such as the Tokyo ETS, covering the commercial and residential buildings sector, can be effective – but also large regional trading schemes such as the EU ETS, which covers electricity generation and energy intensive industries.

Carbon markets as a mechanism for governing climate are fiercely resisted by others, including a wide range of civil society actors and policy think-tanks from both the political Left and political Right (Pearse & Böhm, 2014). Carbon markets are criticised for failing to incentivise the structural changes in the political energy economy that are believed necessary for deep decarbonisation. And for those advocating for radical transformation (Chapter 5), carbon markets are seen as perpetuating climate injustice and development inequalities. They are criticised for propagating a post-political myth that reducing carbon emissions is merely a technical-economic challenge that can be met without confronting the power of incumbent fossil fuel interests. And to this critique one might also add the following. Despite their 15-year operational history and the growing number of carbon markets worldwide, global emissions of carbon dioxide have risen every year, with the exceptions of 2009 (the financial crash) and 2020 (the SARS-CoV-2 virus).[4]

Citizens' assemblies

Concern about the repercussions of climate change has catalysed many social movements over the past quarter century (Chapter 5) and animated siren subaltern voices (Chapter 6). More recently, it has also prompted widespread declarations of "climate emergency" (see Box 9.3). Such civic activism around climate change can be understood as a form of climate governance "from below". As argued by Jason Chilvers and Matthew Kearnes in *Remaking Participation: Science, Environment and Emergent Publics* (2015), climate activism is an example of "uninvited participation" in public life. For example, one of the demands of Extinction Rebellion (XR) in 2019 to the UK government was to establish democratic citizens' assemblies to deliberate on the means for achieving national climate policy objectives. Citizens' assemblies are lauded by deliberative democrats as a way of resolving intractable political issues where it has been difficult to previously reach agreement between political parties.

Citizens' climate assemblies usually function within national or sub-national political jurisdictions and recent years have seen them established in Ireland, the UK, and in France. Neither of these latter two initiatives would likely have been advanced were it not for forceful civic activism on the streets of London and Paris. Partly in response to XR, the UK Parliament established a citizens' assembly during 2020 to consider how the UK can meet the government's legally binding target to reduce greenhouse gas emissions to net-zero by 2050. The policy interventions required to achieve this goal will have implications across the whole society and economy: How homes are heated; what citizens buy; how and where they travel. Citizen participants for "Climate Assembly UK: The Path To Net Zero" were selected randomly from different walks of life, different shades of opinion, and from throughout the

UK. They form a representative sample of the UK's population, a so-called "mini-public" (Smith & Setälä, 2018). Although the results of these deliberations were fed back to Parliament for further consideration, their status was merely advisory. Their recommendations carried no statutory force.

A slightly different process was followed in France. Following the street protests by the *Gilets jaunes* in 2018 and 2019, President Macron established the *Convention Citoyenne pour le Climat* ("the Citizen's Convention on Climate"). This French citizens' convention was given the mandate to define policy measures that would achieve a reduction of at least 40 per cent in greenhouse gas emissions by 2030 (compared to 1990). Importantly – and reflecting the *Gilets jaunes'* protests – these measures were to be defined by the 150 selected French citizens "in a spirit of social justice". Unlike the UK, Macron was committed to executing these citizen proposals "without a filter", submitting them either to a national referendum, a vote in Parliament, or direct regulatory implementation. These examples of citizens' assemblies reveal the generative power of the idea of climate change. It is generative not only of new technologies or pricing mechanisms but also of new forms of democratic participation.

BOX 9.3 GOVERNING IN EMERGENCIES

Each year, the Oxford English Dictionary selects a "Word of the Year". Its designation is intended to highlight "a word or expression shown through usage evidence to reflect the ethos, mood or preoccupations of the passing year". In 2019, the selected Word was "climate emergency". According to the Oxford Corpus – a database containing hundreds of millions of words of written English – the use of climate emergency during 2019 increased a hundredfold compared to the previous year. This exponential rise in public usage reflects a growing trend for political jurisdictions and civic organisations around the world to declare "climate emergencies". The first reported case was in December 2016 by Darebin local council in north Melbourne, Australia. As of January 2021, the Climate Emergency Declaration and Mobilisation in Action website estimated that over 1,850 jurisdictions or organisations had declared a "climate emergency". The declaring bodies were located in 30 different countries and encompassed about 820 million people, just over 10 per cent of the world's population.

These unevenly distributed declarations of climate emergencies are of great interest to geographers. Which bodies are making these declarations and what are their underlying motives? How does declaring a "climate emergency" alter the governance of climate change? Declarations of emergency might convey heightened levels of urgency for responding to climate challenges. They might have the intention of creating a social tipping point to facilitate radical policy measures. But it is not yet clear what the effects of declaring climate emergencies will have on wider climate politics. Declaring an emergency carries many attendant risks. Emergencies create "states of exception", often justified by governments under conditions of war, insurrection, terrorist threat, or, as seen recently, global pandemic (Calhoun, 2008). They promise the mass mobilization of a jurisdiction's full economic, social, and technical capacities to ward off an existential threat.

Yet at the same time emergencies can threaten constitutional rights, everyday freedoms, and can justify the suspension of normal politics. They enable coercive rhetoric, shallow thinking and deliberation, and calls for "desperate measures". Emergencies can also extend state surveillance and control, as was seen with the 9/11 and SARS-CoV-2 emergencies. Declaring a climate emergency is seen by some to legitimise the deployment of "emergency technologies" that might arrest climate change – such as solar geoengineering techniques – by passing statutory democratic processes of consultation and deliberation (Sillmann et al., 2015). One consequence of political declarations of emergency is that the goals of public policy become focused on a limited and reduced set of indicators. For some scholars, meeting the challenge of climate change for future well-being demands a proliferation of diverse policy goals, the very opposite of what "states of exception" bring into being.

These various dangers would all caution against a casual and unthinking embrace of climate emergency. Emergency declarations can be used by both liberal and authoritarian regimes to reproduce "existing and new forms, practices and relations of power" (Anderson et al., 2020: 623).

Read:

Calhoun, C. (2008) A world of emergencies: fear, intervention and the limits of cosmopolitan order. *Canadian Review of Sociology*. 41(4): 373–395; Sillmann, J., Lenton, T., Levermann, A. and co-authors. (2015) Climate emergency – no argument for climate engineering. *Nature Climate Change*. 5(4): 290–292; Anderson, B., Grove, K., Rickards, L. and Kearnes, M. (2020) Slow emergencies: Temporality and the racialized biopolitics of emergency governance. *Progress in Human Geography*. 44(4): 621–639.

Judicial courts

Individuals and social movements take their arguments for participating directly in climate governance to the streets and to citizens' assemblies. But they also may take their appeals directly to the courts. Climate lawsuits brought to judicial process may address either the causes or the consequences of climate change. Thus, the courts may be appealed to in order to prevent public or private actors embarking on new infrastructure projects deemed to damage the climate, such as the expansion of airports or new coal-fired power stations. Alternatively, courts may be appealed to under tort law. Here a claimant seeks compensation for suffering loss or harm as a result of careless or deliberate actions by a third party. The legal entity – often in this case fossil fuel companies – may potentially be found liable for committing "a tortious act", such as wilfully emitting greenhouse gases into the atmosphere.

Over the last 20 years, courts have begun to play an increasing role in the governance of climate change in several countries, as evidenced by the growing number of academic articles dealing with the topic (Figure 9.4). One explanation for this trend has been the perceived failure of international treaties and protocols – and state-sponsored policy measures – to provide adequate governance of the global climate. However, it can also be argued that the Paris Agreement has provided a further impetus for individuals and NGOs to bring their cases to court. Through the mechanism of NDCs, the

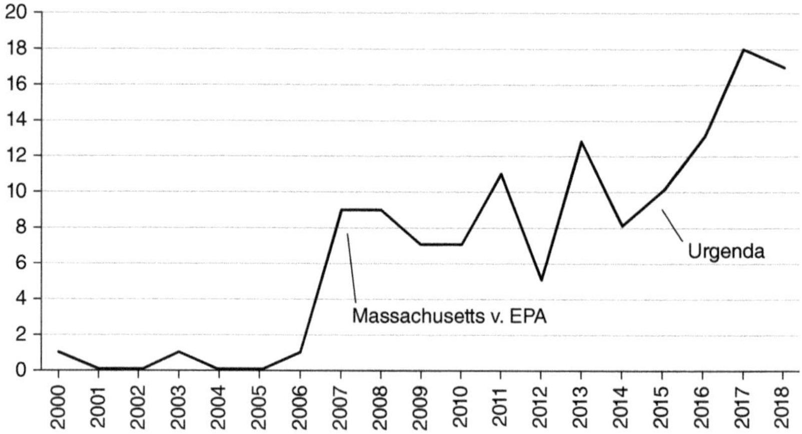

Figure 9.4 The number of academic journal publications per year dealing with climate litigation. Two high-profile lawsuits are marked: *Massachusetts v. EPA* in the USA ruled in 2007; *Urgenda v. The Netherlands* issued its first District Court ruling in 2015.

(*Source*: Setzer & Vanhala, 2019).

agreement acknowledges the primacy of national policy making over international orchestration. This offers new motivation for litigators to use judicial processes to hold national governments and other actors to legal account.

One of the most salient legal climate change cases has been that of *Urgenda Foundation v The Netherlands* (2015). A Dutch NGO – the Urgenda Foundation – brought a summons against the Dutch government in 2013 under the Dutch civil law doctrine of "hazardous state negligence". Urgenda claimed that the government's climate policy measures designed to reduce emissions of greenhouse gases were insufficiently ambitious in relation to the hazard presented to the Dutch people by climate change. After nearly five years of hearings and appeals, in December 2019 the Supreme Court in The Hague finally ruled in favour of the plaintiffs. The court ordered the Dutch Government to reduce national emissions by 2020 by 25 per cent relative to 1990 levels. Interestingly, this had in fact always been the stated goal of successive governments, but the *Urgenda* ruling – by making this reduction a *legal* requirement – moved the locus of climate governance from a democratically elected parliament to unelected judges in the courtroom. A useful discussion about the merits and demerits of activating the courts in climate governance can be found in Scotford et al. (2020). The significance of the *Urgenda* case is that the state – or more precisely the Dutch state – is now subject to the law when setting climate policy.

Climate services

Enhancing the adaptive capacity of societies was introduced earlier in the chapter as one possible object of climate governance. One mechanism by which such governance can be advanced is through the institutional innovation of "climate services". Climate services aim "to provide people and organizations with timely, tailored climate-related knowledge and information that they can use to reduce climate-related losses and enhance benefits, including the protection of lives, livelihoods, and property" (Vaughan & Dessai, 2014: 588). One example of the tools being developed for this purpose is the "Forecast in Context Map Room", developed by the International Federation of Red Cross and Red Crescent Societies to assist in disaster-related decision making. This tool presents regularly updated six-day and three-month forecasts of rainfall, calibrated against local averages, to help disaster risk managers identify early intervention actions based on given forecasts. Climate services such as this have expanded greatly in recent years, stimulated by the adoption by the World Meteorological Organisation in 2012 of the Global Framework for Climate Services.

Many humanitarian organizations, government offices, international agencies, and the private sector have sought out these new climate services as a way to improve climate risk management and governance. These service providers now exist at local, national, regional, and international scales. They offer information and management services to a range of different sectors – agriculture, health, forestry, fisheries, transport, tourism, disaster risk reduction, water resources management, and energy. The rapid commercialisation of such service provision raises concerns, however, about the privatisation of adaptation governance. Under such a privatised business model, access to climate services will be geographically uneven, leading to inequalities in improved societal resilience to climatic hazards that will likely mirror existing economic inequalities (Webber & Donner, 2017).

Chapter summary

Environmental historians such as Sara Miglietti have been able to show that projects to govern the climate have always ended up being about more than controlling climate, even if they were not originally designed to do so. For example, in *Governing the Environment in the Early Modern World: Theory and Practice* (2017), Miglietti and John Morgan offer a valuable historical perspective on the contemporary politics of climate knowledge and the conundrums of climate governance. In a series of historical case studies, they take their readers back to the marshes of seventeenth-century France, the imperial Arctic projects of early nineteenth-century Britain, the wetlands of the newly independent east coast states of North America, and the sixteenth-century Spanish colonies of South America. These cases explore how past societies have sought to govern their environments. Governing local or colonial *climates* in the past was always an exercise in governing local *communities* or colonial *societies*.

In similar fashion, the project pursued over recent decades of global climate governance has meant controlling things other than climate. The headline object of climate governance as specified in the Paris Agreement is global temperature. But as this chapter has shown, global temperature is not an entity that is directly tractable to intentional human action. Governing temperature therefore requires governing the full range of human activities and technologies that emit greenhouse gases and particulates into the atmosphere and the imaginations that give rise to them. This requires virtually every human practice becoming subject, at least in principle, to the logic of climate governance. Diet, energy, forests, land, materials, mobility, recreation – ultimately all human behaviour, even potentially procreation – become subject to its totalising reach. Climate governance also extends to enhancing the adaptive capacity of societies to better protect themselves against climatic hazards.

Governing global climate therefore becomes an exercise in governing the collective of human societies but where the power to do so exists in no central or identifiable location. This is one of the reasons why the locus of climate governance has moved decisively away from the orchestration of nation-state actions through the UN, to instead be distributed more widely across an expanding multi-scalar matrix of diverse social actors and regulatory regimes. Far from Biermann's vision of a coordinated and intelligent Earth System Governance framework, a more plausible metaphor for climate governance is that of a clumsy multi-layered meshwork of overlapping and competing competences and interests. History suggests that most projects of centralised vision and control are doomed to become fantasies that fail. As James Scott pointed out more than 20 years ago (Scott, 1998), social, environmental, and cultural contingencies always overwhelm the ordering eye of power. The human instinct and desire for controlled and predictable outcomes will always be thwarted by events, political conflict, and by people's own individual and collective fallibility.

Notes

1 It is worth noting that "global temperature" is not an easily defined or measured quantity. Scientifically, it is taken to be the globally averaged near-surface mean air temperature of the planet. A useful discussion of how this calculation is made – and its uncertainties – can be found in Jones et al. (1999).

2 Neither is fertility control mentioned directly in any of the 17 SDG goals or associated 169 SDG targets, even though a number of these targets – such as increasing female education and universal access to reproductive health services – will reduce fertility rates. Previous political, ethical, and religious disputes about international "population policies" continue to require the goal of fertility reduction to be approached indirectly in intergovernmental settings.

3 The six countries not yet ratified are Eritrea, Iran, Iraq, Libya, Turkey, and Yemen. The USA gave notice to withdraw under President Trump, a decision reversed in 2021 by President Biden.

4 It is worth noting that these two exceptions would suggest that reducing the overall level of activity in the global economy is a more effective way of reducing carbon emissions than carbon pricing – but a reduction that comes at considerable cost to human employment, security, and welfare.

Further Reading

The best reference source on climate governance is **Karin Bäckstrand** and **Eva Lövbrand's** edited *Research Handbook on Climate Governance* (Edward Elgar, 2015), which contains 50 essays on all aspects of climate governance. A good historical overview of the evolution of global climate governance from the 1980s onwards is **Joyeeta Gupta's** *The History of Global Climate Governance* (Cambridge University Press, 2014). In *Governing Climate Change: Polycentricity in Action?* (Cambridge University Press, 2018), **Andrew Jordan, Dave Huitema, Haro van Asselt,** and **Johanna Forster** edit a series of essays that evaluate the achievements and potentials of polycentric climate governance. Also in the line of polycentric governance is *Governing Climate*

Change: Global Cities and Transnational Lawmaking (Cambridge University Press, 2018), in which **Jolene Lin** examines the agency of city authorities to act as catalysts and network nodes for climate governance. In contrast to polycentrism, **Anatol Lieven** argues powerfully in *Climate Change and the Nation State: The Realist Case* (Allen Lane, 2020) for a state-centred, even nationalist, strategy for tackling climate change. For a good exploration of the role of the private sector in climate governance and why it matters, read **Michael Vendenbergh's** and **Jonathan Gilligan's** *Beyond Politics: The Private Governance Response to Climate Change* (Cambridge University Press, 2017). And more specifically on the role of emissions trading and carbon markets as a form of climate governance, **Janelle Knox-Hayes's** *The Cultures of Markets: The Political Economy of Climate Governance* (Oxford University Press, 2016) is to be recommended. In *Accomplishing Climate Governance* (Cambridge University Press, 2016), **Harriet Bulkeley** offers an original and persuasive account of the everyday politics and geographies of climate governance using the example of the UK. To explore questions about how solar geoengineering might be governed, then **Jesse Reynolds'** *The Governance of Solar Geoengineering: Managing Climate Change in the Anthropocene* (Cambridge University Press, 2019) offers a useful, though perhaps overly optimistic, starting point. In *Contemporary Climate Change Debates: A Student Primer* (Abingdon: Routledge), **Mike Hulme** has brought together in an introductory text many of the debates and points of tension about how climate is and should be governed using the device of twinned essays from invited scholars arguing for and against specific propositions.

QUESTIONS FOR CLASS DISCUSSION OR ASSESSMENT

Q9.1: How essential are *global* numerical targets for the effective implementation of national and sectoral climate policies?

Q9.2: Are there risks in relying more on nonstate rather than on state authorities to deliver climate change mitigation? How might these risks be alleviated?

Q9.3: Do you think that national – and international – courts should play a larger role in holding polluting industries to account? What about for holding national governments accountable? What are some of the risks of using courts in this way?

Q9.4: What difference does it make to the design and execution of climate policies to declare climate change "an emergency"? Is this a good thing?

Q9.5: In what ways – if any – should concerns about climate change have bearing on your decision to procreate?

Q9.6: To what extent do you believe global climate is something that *can* be governed?

10

CLIMATE IMAGINARIES

. . . Climate change forever . . .

Introduction

The future of climate is underdetermined. That is to say, rather obviously, that the climatic future has not yet arrived. But it is also to say that the future of climate is not known – scientific prediction, climate governance, and religious or cultural prophecy notwithstanding. Or at least the future is not known with any certainty. But if the climatic future has not yet arrived and is not yet known, it can certainly be imagined. Historically, and perhaps even more so today, future *climate imaginaries* – see Box 10.1 for a definition – wield extraordinary power over the present. This chapter sets out to explain how future climates are imagined and by whom. In doing so, it relates some of these imaginaries to the different narrative positions and meanings of climate change outlined in preceding chapters.

The desire to know the future – not least the desire to discern future changes in climate – has been a source of cultural innovation throughout human history. At moments of great uncertainty or crisis, a community or society will turn to its prophets in search of guidance – for clarity about how the future will evolve, for insight about how to act, for assurance to assuage fears and anxieties (Walsh, 2013). In George Orwell's haunting novel *Nineteen Eighty Four* (Orwell, 1949), his protagonist Winston Smith, observing the hegemony of the totalitarian party "Ingsoc", comes to realise that "who controls the past, controls the future: who controls the present, controls the past". In Orwell's imaginary superstate of Oceania, Ingsoc wields complete political power in the present, thus enabling it to control the way in which its subjects think about and interpret the past. And such control of the past ensures control of the future. The inhabitants of Oceania are conditioned to pursue the party's future goals because Ingsoc ensures that the past is told in a way that justifies and incentivises them.

With respect to climate change, Orwell's aphorism might be extended full circle: "Who controls the future, controls the present". Whichever imaginative account of climate's future gains powerful ascendancy in today's world – whether that power be political, cultural, or epistemic – positions itself as a disciplining force over human subjects in the present. Another way of saying this is that persuasive stories of approaching climate change exert remarkable, almost mesmeric, power over the human imagination. The environmental historian Lucien Boia has elucidated this phenomenon historically in his book *The Weather in the Imagination* (2005). And an example of this power being exercised in the present is David Wallace-Wells' portrait of future climatic devastation in *The Uninhabitable Earth* (2019), which gained widespread media and popular attention in the English-speaking world.

To explain this power of a future climate imaginary I turn to German sociologist Ulrich Beck and his influential *Risikogesellschaft* ("risk society") thesis of the 1980s. Beck proposed this notion of risk society to describe societies increasingly preoccupied with future risk and hence reorganising in new ways to deal with the hazards and insecurities introduced by modernity. Beck's modernist theory of risk aptly applies to the idea of climate change. In particular, he recognises the power exerted over the present by the shadow cast by imagined futures. "[Climate] risk is not the same as catastrophe", says Beck, "but [risk is] the *anticipation* of the future catastrophe in the present. As a result, [climate] risk leads a dubious, insidious, would-be, fictitious, allusive existence: it is existent and non-existent, present and absent, doubtful and real" (Beck, 2009: 4; emphasis added). Beck is no climate sceptic or contrarian; physical climates are indeed changing. But he points to the ability of future imaginaries to exert influence on today's world. Events that have not yet happened in reality, happen in the imagination. "Expected risks are the whip to keep the present in line", Beck observes. "The more threatening the shadows that fall on the present because a terrible future is impending, the more believed are the headlines provoked by the dramatisation of risk today" (Beck, 1997: 20). Thus, it is demonstrated: Who controls the future, controls the present.

It is for this reason that the three narrative positions on climate change outlined in Chapters 3, 4, and 5 each sought to ground their manifestos for political action – or inaction – in their own interpretations of the climatic future predicted by climate science. It is why the "more-than-science" approaches for sense-making around climate change outlined in Chapters 6, 7, and 8 each offered their own imaginaries of the climatic future. And it is why future climate imaginaries, translated into ever-shortening deadlines within which to enact social change, are used to govern the present in the name of climate change (Chapter 9). Futures specialist Renate Tyszczuk and colleagues (2019: 10) describe in this way the mesmeric power possessed by future climate:

> [Climate change] is a scenario that simultaneously conjures up visions of an uninhabitable earth, unimaginable societal transformations, future extinctions and expulsions. It summons authoritarian technofixes and existential crises. It backtracks and fastforwards searching for patterns and trajectories. It tracks the convulsions of a restless earth and proposes monstrous solutions. In different hands it becomes a mandate for taking back control (of the planet), or an argument for giving up, or a motive for trying to leave entirely. Yet climate change as an idea is at the same time an invocation – a holding out hope for a better world. It wants to think the future otherwise. It has inspired a multitude of speculative fixes – Plans B through Z, and beyond.

If this power of the imagined climatic future over the present is indeed real then different questions need asking beyond the scientist's simple, "What will happen to climate?" And they are important questions. Who controls the idea of climate change as it encounters the future? Why and how are different future climate imaginaries sustained? Are these climate futures global or local in scale? The appropriation of climate science by diverse political actors, a process elaborated in Chapters 3, 4, and 5, might suggest that it is scientific claims about future global climate that exert the greatest power in the present. But is this really so? There are those for whom it is a priority to listen to subaltern voices and *their* claims over the climatic future emanating from varied places, alternative ontologies and political expressions of resistance (Chapter 6). There are others for whom the re-imaginings of future climates offered by creative artists are powerful and provocative (Chapter 7). And many may be drawn to stories about climate's future emanating from religious imaginations (Chapter 8). The hopeful imaginary offered up in the Paris Agreement, of a global climate future no warmer than between 1.5° and 2°C above the nineteenth-century baseline (Chapter 9), does not necessarily trump all other climate imaginaries.

Seen this way, the exploration of the idea of climate change in the past and present, completed in earlier chapters, has been a quest to reveal different climate imaginaries circulating in different social worlds. Some of these imaginaries are more rooted in science; others less so. It prompts the obvious question: Whose imaginary counts most?

***

There are many different cultural practices for imagining the future and this chapter will investigate a few of them with respect to future climates. These will resonate in various ways with some of the epistemic claims and representational practices and politics encountered in earlier chapters. Scientific predictions are of course one such mode of *futuring*, a systematic process for thinking about the future, picturing possible outcomes and planning for them. But as has been made clear in earlier chapters, claims to speak for climate's future do not emanate from science alone. Other futuring practices include scenarios, speculative fiction, prophecies, divinations, visualisations, metaphors, and myths. These practices lead to a range of climate imaginaries beyond those that could possibly emerge either from the scientist's remit to predict or from the IPCC's remit to assess scientific knowledge. Full disclosure of the multiple meanings of climate change requires the adoption of "more-than-science" approaches for sense-making. So too does a full exploration of future climate imaginaries.

There are two further important observations to make at this point. Just as imagining the climatic future cannot be left to *science* alone, so imagining the future cannot be reduced to *climate* alone. The temptation to do so is great. The phenomenon of *climate reductionism* results from the following dubious logic: If social change is unpredictable but climate change predictable then the future can be made known by elevating climate as the primary driver of social change. (Climate reductionism – and its relationship with climatic determinism, see Box 1.1 – is explained in Hulme, 2011.) But the planetary future is not conditioned on climate alone, nor just on physical ecologies. The future is many-sided and far from being solely climate-shaped. It is also about forthcoming social, political, economic, cultural, medical, technological, and religious changes that are, to different degrees, largely unforeseeable. Not least this is because of **Black Swan events** such as terrorist outrages (think 9/11) or pandemics (think SARS-CoV-2). This many-sided future is why the scope of climate governance keeps expanding, as we saw in Chapter 9. One cannot govern climate without governing the future. And one cannot govern the future without governing everything, including the imagination and its dystopias and utopias. Climate's future is conjoined with that of humanity, both materially and imaginatively.

The second observation is about the geographies of futuring. If there is a temptation to reduce the future to climate, there is also a temptation to default to climate futures that are ungrounded in specific places and cultures. Imagined future climates are often portrayed at a global scale, not least because of the inspiration taken from global climate models and their futurist predictions. But climate futures may also emerge from locations other than rootless numerical models and through practices and imaginaries grounded in specific histories and cultures. There is an important difference between singular global climate futures and futures featuring compositions of multiple local climates. A *global* perspective on the future is not the only option.

In surveying climate's imagined futures, the chapter draws on *futures studies*, *science and technology studies*, and *ecocriticism*. It explains how the climatic future is imagined and what sorts of futures are being imagined. In the first part of the chapter four futuring practices are introduced: *Scenario planning* and *scientific modelling*, both of which can be thought of as broadly realist techniques and *metaphors* and *creative fiction*, both of which offer more speculative modes of futuring. Climate imaginaries of the future frequently revert to the archetypes of dystopias and utopias, of which a number of examples are given. The distinctions between dystopic and utopic climate futures, and the political struggles to realise them, are nowhere better illustrated than through the imaginary of geoengineered technoclimates and these are briefly discussed. The chapter concludes with some vignettes of the climatic future emerging from different imaginaries. Climate change will forever be in the human imagination as much as it shall continue to reshape social and physical worlds.

Futuring practices

This subsection elaborates four *futuring practices* as applied to the climatic future, four creative ways in which groups of people apply their imaginative faculties and cultural resources to engage the future. Two of these can be thought of as offering realist techniques of futuring: Scenario planning and scientific modelling. They seek to develop plausible and rational accounts of what the climatic future might look like. The other two techniques considered here – metaphors and creative fiction – should be thought of as speculative practices. They offer images and stories that provoke the imagination about the climatic future. All four practices are capable of bringing into being the full variety of climate futures described earlier by Renate Tyszczuk – the benign, violent, fearful, censured, utopic, or dystopic. Futuring practices are closely linked to the idea of "social imaginaries" and "sociotechnical imaginaries" (see Box 10.1).

BOX 10.1 CLIMATE IMAGINARIES

Within sociology, the concept of "the social imaginary" embraces a set of values, institutions, laws, discourses, and symbols through which people imagine themselves as a social whole in a wider world. Imaginaries can be thought of as ways in which societies make sense of their environment and their place in the world. Although the modern concept of the imaginary can be traced back to Jean-Paul Sartre in the 1940s, the Canadian philosopher Charles Taylor gave it recent and wider salience through his book *Modern Social Imaginaries* (2003). For

(Continued)

Taylor, social imaginaries are semiotic systems that give meaning to events; they shape the practices, lived experiences, and identities of social groups. And importantly for this chapter, social imaginaries can also provide orientation towards the future.

Imaginaries must be understood as more than individually held beliefs or assumptions. To be stable and functional, an imaginary must be held in common by the members of a particular social group. They are always historically and geographically situated. They may be globally or locally oriented. Social imaginaries are also interlinked with material forms of "world-making", notably scientific research and technological invention. This point has been most clearly articulated by Sheila Jasanoff and Sang-Hyun Kim in *Dreamscapes of Modernity* (2015), in which they argue that social imaginaries are coproduced with scientific and technological developments. This leads them to propose the notion of *sociotechnical imaginaries*. By this they mean that whilst technologies and scientific findings – for example predictions about future global climate – influence and shape how people imagine their environment, conversely social imaginaries shape the path of scientific and technological development. As is also made clear by Jasanoff and Kim, imaginaries are not *merely* imaginaries. They are not simply inert figments of a fertile imagination. Sociotechnical imaginaries operate across the boundaries of the perceptual *and* the material. They can bring real worlds into being, for example carbon capture technologies, driverless vehicles, intelligent robots, or space tourism.

Applying this understanding of "the imaginary" to climate change means that *climate imaginaries* can be understood as collectively shared sets of beliefs, narratives, technologies, discourses, and practices that condition what climate futures are thought of as possible, likely, or (un)desirable. Climate imaginaries envision not only possible climate futures, brought to life through different futuring practices. They are also suggestive of ways to deliver such futures. For this reason, all imaginaries are politically charged.

Read:

Taylor, C. (2003) *Modern Social Imaginaries*. Durham, NC: Duke University Press; Jasanoff, S. and Kim, S.H. (eds.) (2015) *Dreamscapes of Modernity: Sociotechnical Imaginaries and the Fabrication of Power*. Chicago: The University of Chicago Press; Benner, A-K., Rothe, D., Ullström, S. and and Stripple, J. (2019) *Violent Climate Imaginaries: Science-Fiction-Politics*. Hamburg: Institute for Peace Research and Security Policy.

Scenario thinking

The origin of "scenarios" lies in the world of theatre, in particular sixteenth-century Italian street theatre (Tyszczuk, 2019). Scenarios were first articulated as part of an improvised form of theatre known in Italy as *commedia dell'arte* – literally "comedy of the profession". The unscripted "scenario" comprised the skeletal framework of a story intended to be fleshed out through improvised acting. Scenarios, whilst offering a loose narrative structure for the performance, left much room for the speculative and the surprising.

This idea of a scenario – a rough outline of a playtext to be elaborated through improvisation – travelled from Italian street theatre, via the scripts of Hollywood screenplays, into the strategic military thinking of the USA during the Cold War and the multinational corporate boardrooms of the 1950s and 1960s. Scenario planning today is widely recognized as a formalised technique for managing high levels of uncertainty when determining future business strategies and government policies. It is also integral to video games in city-building and turn-based strategy genres such as *Civilisation VI* and *Frostpunk*, both of which have climate scenarios embedded within them. Scenario thinking underpinned the new environmental modelling contained in the *Limits to Growth* reports of the 1970s (Meadows et al., 1972), while the first "climate change scenario" appeared in the academic literature in 1977 (Flohn, 1977). In all of these forward-facing settings, the purpose of a scenario is to invite creative thought about different possible futures. As befits their theatrical origin, scenarios stimulate collective, improvised, and reflexive ways of thinking about uncertain futures.

Scenario techniques were adopted by the IPCC during the 1990s and 2000s and some of these climate scenario practices were discussed in Chapter 2. For example, using scenario methods, the IPCC created future socio-development pathways and greenhouse gas emissions trajectories. In these activities, the IPCC inherited a structuring device common in scenario design, namely the two-by-two matrix – sometimes referred to as "the Boston Matrix".[1] The device is simple. Two contrasting axes are created, which are used to structure the scenario analyst's subjective judgements about the future. They might be, for example, a "centralisation–decentralisation" axis versus an "individualism–communalism" axis, if thinking about governance futures. Or if thinking about ecological futures, a "biotech–natural" axis versus an "intrinsic–extrinsic values" axis. The result of such differentiation is the elaboration of four scenarios according to the four resulting quadrants (see Box 10.2 for an example of this). This scenario device has been widely used by institutions with regard to climate change, ranging from the IPCC to the World Bank and from oil companies to citizens' assemblies.

BOX 10.2 GEOPOLITICAL CLIMATE SCENARIOS

Much scenario thinking around climate change has focused on climate, energy, development, and ecological futures. But North American geographers Joel Wainwright and Geoff Mann applied scenario analysis to explore a range of geopolitical futures that might emerge in response to the governance challenges of climate change. First developed in a 2013 article in the radical geography journal *Antipode*, Wainwright and Mann elaborated the framework at much greater length in a 2018 book *Climate Leviathan: A Political Theory of our Planetary Future*. Following the logic of the Boston Matrix, their scenario thinking posits two axes along which future political systems may organise: A capitalist–socialist axis and a planetary–nationalistic sovereignty axis (Figure 10.1). This led them to propose four geopolitical arrangements that may emerge in response to climate change, namely:

	Planetary Sovereignty	Anti-planetary Sovereignty
Capitalist	Climate Leviathan	Climate Behemoth
Non-capitalist	Climate Mao	Climate X

Figure 10.1 Four geopolitical scenarios emerging in response to climate change. (*Source*: Wainwright and Mann, 2013).

Climate Leviathan – a system of global capitalism governed by a planetary sovereign, perhaps some hegemonic power capable of taking decisive and centralised action.
Climate Mao – an anti-capitalist system governed by sovereign powers at the level of the nation-state or regional self-interest.
Climate Behemoth – a capitalist system within the confines of absolutist nation-states.
Climate X – a rejection of both capitalism and state sovereignty for something yet to be determined.

A related example of futures-thinking guided by scenario logic is offered by Morgan Bazilian and colleagues (2020). These authors attended specifically to the future politics of different energy transitions that may emerge in response to climate change. They too identified four different scenarios, but in this case each scenario was conditioned by just one dominant driver of change. Thus, their *Big Green Deal* scenario resulted from concerted multi-lateral policy initiatives; *Dirty Nationalism* from nation-first policies; *Muddling On* from falling energy costs but slow technical progress; and *Tech Breakthrough* from disruptive advancements in energy technology.

Both these examples capture the creative and improvised spirit of Italian street theatre where the idea of a scenario originated. Wainwright and Mann admit their scenarios are somewhat vague and amorphous, but nevertheless they provoke the political imagination. Rather than offering firm predictions or set scripts, these scenarios offer only sketches of possible future political arrangements. Scenarios such as these are offered as imaginative resources – not least through suggestive namings – which different political interests and actors might exploit for strategic advantage. Nevertheless, introducing scenarios to the public domain is one way of structuring – and therefore exerting power over – the unknown future.

Read:

Wainwright, J. and Mann, G. (2013) Climate Leviathan. *Antipode*. 45(1): 1–22; Wainwright, J. and Mann, G. (2018) *Climate Leviathan: A Political Theory of our Planetary Future*. London: Verso Books; Bazilian, M., Bradshaw, M., Gabriel, J., Goldthau, A. and Westphal, K. (2020) Four scenarios of the energy transition: Drivers, consequences, and implications for geopolitics. *WIREs Climate Change*. 11(2): e625.7pp.

Model Predictions

Climate model predictions were encountered in earlier chapters. For example, Chapter 2 explained why numerical models of the climate system emerged during the latter decades of the twentieth century. By offering prognoses of the future, model predictions appear to deliver on science's promise of comprehensive explanation of the physical world. As a futuring practice, climate models have secured extraordinary epistemic authority and cultural privilege in climate change discourse (Hulme, 2013). For example, climate model predictions have been the cornerstone of successive IPCC reports, elaborating with ever greater precision[2] how the climate future might turn out. Importantly, the default setting for such futuring practice is global – not local. But model predictions, as also do scenarios, enter into charged political and social spaces. Heather Houser rightly observes that "the importance of models is incontestable because they are public stages for climate debate and have become knowledge battlegrounds" (Houser, 2020: 34). And as we saw in Chapters 3, 4, and 5, different political positions on climate change find themselves tussling to secure the right to issue authoritative interpretations of climate model predictions.

Analysed as a futuring practice, climate model predictions offer something between, on the one hand, all-knowing revelations of the future and, on the other, socially constructed myths concerning the future. The epistemic foundation upon which models predict future climates comprises a combination of physical theory, observable reality, creative speculation, and numerical simulation. The speculation arises because all model predictions are contingent, to a greater or lesser degree, upon the futures conjured through the types of scenario exercises described earlier (Figure 10.2). Another way of capturing this ambiguity of climate model predictions is to say that they should be taken seriously but not literally.

Climate model predictions impose themselves on the future by claiming, if not certainty, then at least the ability to quantify **uncertainty**. The promise then follows of being about to reduce this uncertainty through future scientific research. But for some scholars this search for certainty about future climate is exactly the opposite of how the future should be approached. The future is inherently uncertain, they would say, the climatic future as much as any other dimension. And so, in *The Politics of Uncertainty: Challenges of Transformation* (2020), Ian Scoones and Andy Stirling argue that rather than embarking upon impossible quests to *eliminate* uncertainties about the climatic future, the real quest should be to *understand* uncertainty's enduring presence and social character. This requires recognising uncertainty as a complex construction of knowledge, materiality, experience, embodiment, and practice. In the same volume, Lyla Mehta and Shilpi Srivastava argue that "subjective judgements, multiple knowledges and

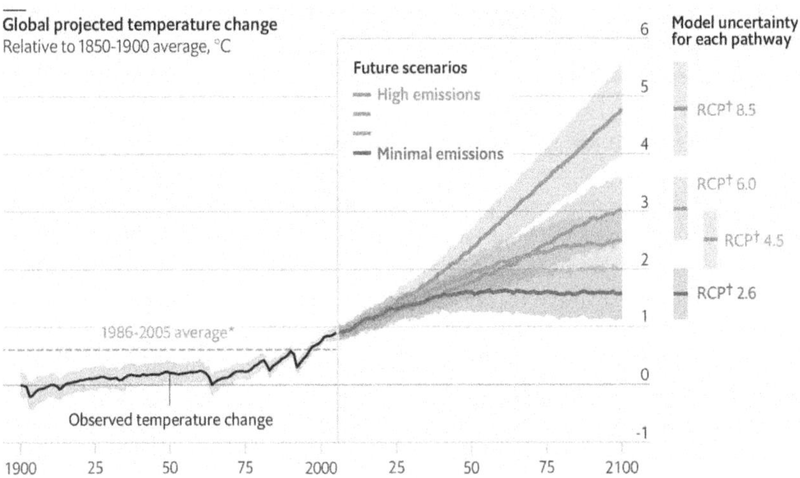

Figure 10.2 Global temperature change relative to 1850–1900 simulated by CMIP5 global climate models for each of the four RCP emissions scenarios.
(*Source*: Knutti, R. and Sedláček, J., 2013).

diverse interpretations around uncertainty are inevitable and must be central to responses to uncertain situations" (Scoones & Stirling, 2020: 100). Uncertainty about future climate depends deeply on political values and social choices, which are shaped by historical and cultural processes. It is not just a technical issue that more research and more powerful models can eliminate.

Metaphors

Scenarios and model predictions offer, to varying degrees, realist and rational means for creating climate futures. The futuring practices of metaphors and speculative fiction provoke more imaginative ways of thinking about future climates. Metaphors are powerful linguistic and visual devices for guiding human cognition, emotion, and behaviour. Metaphors help us grasp something new or unfamiliar by associating it with something more familiar and everyday. For example, one of the earliest metaphors used to communicate the idea of human-caused climate change was that of the greenhouse. Gases in the atmosphere that absorbed outgoing heat were described as "greenhouse gases" and the resulting effect on the climate was described as "the greenhouse effect". Metaphors are of course not intended to be literal – the Earth's atmosphere is *not* a greenhouse. But they help explain an idea, enable a comparison, or provoke a line of thought.

Metaphors are indispensable to both scientific inquiry and literary fiction. Things can never be understood "as they really are" but only through the concepts and language we use to describe them. And metaphors are essential in this task. Ecologist Brendon Larson, in *Metaphors for Environmental Sustainability* (2011), emphasises the importance of choosing appropriate metaphors through which to communicate. Different metaphorical choices carry different ethical or political implications. One of his examples applied to ecosystems applies equally to climate. He points out the significant difference between using the metaphors of either "restoration" or "engineering" to communicate the objectives of environmental policy. Restoration implies the possibility of reverting an ecosystem to some prior, "natural" condition; engineering implies that there is some degree of enhancement or artifice involved.

Because metaphors offer a powerful medium through which to frame political action, they should be considered as powerful devices for thinking about the climatic future. In subtle or not so subtle ways, metaphors shape human cognition and behaviour. They are part of the imaginative structures that are shaping future worlds. "Through the entanglements of social, technological, and natural agencies", observe Maggs and Robinson (2020: 11), "the real world is produced by the imaginary structures of human culture as never before". Take, for example, the phrases "tipping points", "planetary boundaries", "risk cascades" and "runaway climate change". These expressions, emanating from Earth System scientists, have become commonplace in everyday discourse and political speech about future climate. They again project an imaginary that is global rather than local in scale. But it is important to recognise them as metaphors, as imaginative ways of thinking about and communicating scientific knowledge, rather than as literal descriptions of things "as they really are" or the future "as it really is".

This communicative function is what metaphors are good for. But they are not neutral with respect to their effects on the human imagination or on political action. Consider another set of metaphors that are used to communicate the idea of deliberately altering the Earth's heat balance through injecting aerosols into the stratosphere. This presumed technology has been variously described as creating a "global thermostat" for the planet, or as offering Earth a "sunscreen" or presenting humanity with an "insurance policy" (Nerlich & Jaspal, 2012). These different metaphors for solar geoengineering condition audiences to think of these global climate control technologies in mechanical, medical, or economic terms.

Metaphors can be hard to spot and can act as political Trojan horses. Take the notion of the global carbon budget. Over the last decade this has become an established framework for thinking about the design and execution of climate policy, not least as articulated in the Paris Agreement. But the metaphor of "a budget" is strongly connotative of financial management and suggestive of associated metaphors such as deficits, overdrafts, loans, bankruptcy, and so on. These metaphors all resonate with the market environmentalism favoured by neoliberal ideology. The metaphor of a "carbon budget" may appear neutral and communicatively useful, but its associations with fiscal instruments of neoliberalism easily pass unnoticed. Such a metaphor subtly shapes and conditions ways of thinking about the goals and techniques of climate management and policy.

The label "**The Anthropocene**" might also be understood as a metaphor. A literal translation of Anthropocene is "the age of humans", a meaning that immediately is suggestive of human power and ascendancy. But the metaphor can be read in different ways (Rickards, 2015). It points to a tension. This is between human agency and its manipulative power over the material world on the one hand and the enduring non-human agency that resides in oceans, glaciers, storms, species, viruses, and genes on the other. When deployed in relation to the climatic future, the Anthropocene provokes powerful socio-technical imaginaries. Is the Anthropocene a way of drawing attention to the awesome – but unequal – powers and responsibilities people now have for shaping the climatic future? Does it provoke a questioning of the character and wisdom of the Anthropos – the human – who has given rise to this epoch and its unequal power relations? Or does the Anthropocene metaphor dissolve the old binaries of modernity that separate nature from culture and so recognises that climate is no longer natural and never again can be?

Speculative fiction

In Chapter 7, I argued that speculative climate fiction allows authors many more degrees of freedom to engage the climatic imagination than is the case with realist literature. This partly explains the prevalence of this genre of writing in relation to climate change. Such writing brings into being future worlds in which people are struggling to survive, adapt to, or mitigate climate change. As we shall see in the next subsection, many of these speculative climates are associated with dystopic and often violent futures. Only a few paint more optimistic climate futures. But the central point for this discussion is the powerful ability of literature to conjure future climates into being through the reader's imagination.

Such speculative climate futures should not be thought of as prescriptive or normative. And they certainly rest on different epistemic foundations than the prognostications of climate scenarios or model predictions. And they do not necessarily default to the global scale, as tends to be the case with climate modelling. As with metaphors, speculative fiction is intended to provoke the imagination – but perhaps in more explicit and self-conscious ways. Such fiction offers subjective accounts – fictional stories – of how people of the future might interact with climates of the future. These stories are often grounded in specific places and speak directly to the reader's emotional and political imagination. Is this climatic future one in which I wish to dwell? Does this story place any ethical or moral obligations on me to act in certain ways in today's world? Do I learn anything about myself from the way in which these characters live out their lives in a climate-changed future world? Powerful speculative fiction may not in any direct material sense change the course of future climate. But it may address what Ghosh (2016) diagnoses as a "lack of imagination" in current climate change discourse about the possibilities of change.

One interesting foray into speculative climate fiction was made by science historians Naomi Oreskes and Erik Conway in *The Collapse of Western Civilization: A View from The Future* (2014). This is an unusual piece of writing, not least because historians are not known for telling stories about the future. Historians are much more interested in constructing stories about the past, where historical facts are woven together to provide convincing accounts of "how one thing led to another". But there are no "facts" about the future for historians to discover or construct. As we have seen, the future is usually left to the imagination of novelists, the secret knowledge of seers, or the predictive models of scientists. Oreskes and Conway's story is about the collapse of Western civilization as told by a Chinese historian in 2393 CE. They write from a doubly privileged stance. First, they create a fictional future climate. And then, writing as an imaginary historian living on the "other side" of this created future, they pass judgement on the actions of those living today who created this future they have conjured into being. *The Collapse of Western Civilization* is an unusual form of speculative fiction, perhaps best characterized as prophecy written as history.

Climate dystopias/utopias

The various futuring practices described above give rise to a wide range of climate imaginaries that combine numerical, visual, narrative, and symbolic forms of representing the future. Nevertheless, there seems an enduring tension in public climate change discourse between two archetypical climate imaginaries. On the one hand is the deep pessimism expressed in visions of global climate breakdown and civilizational collapse, of which Wallace-Wells' *The Uninhabitable Earth* mentioned earlier is emblematic. These examples might be thought of as climate dystopias. On the other hand, is the optimism of transformative radicalism (Chapter 5), evidenced for example in the rhetoric of the green new deal and in the visionary hopes of climate social movements. Some of these visions for the future might be classed as climate utopias.

This is not to say that there is a balance or symmetry in the manifestation of these different climate imaginaries. There remains a dominant tendency to view climate's future in dark terms; few people it would seem are instinctively climate optimists. But there is a strange relationship between these two contrasting archetypical climate imaginaries that has been evident throughout many of the earlier chapters of the book. The human imagination seems to engage with climate's future in a way that oscillates between deep despair and the hope of salvation. Despite the prevalence of dystopic climate imaginaries, there remains a surging belief in the possibility of averting these imagined futures. Literary scholar Robert Markley observes the phenomenon thus: "Even as climate fiction . . . veers towards consensual visions of a dystopian future, it tacitly assumes that anthropogenic climate change can be mitigated by heroic actions of managerial expertise" (Markley, 2019: 23). Climate dystopias are needed, it might be said, in order to provide the traction needed for political hope to mobilise.

In this subsection I analyse some of these imaginaries through the contrasting lenses of dystopias and utopias. In framing climate imaginaries in these terms, it is worth noting that, historically, utopias preceded dystopias. Utopias – *ou topos* in Greek, literally "not places" – can be traced back to Thomas More's 1516 eponymous book about a fictional island society. Dystopias on the other hand – *dys topos*, literally "not-good places" – found its originating author in John Stuart Mill only in 1868. In literary terms, utopias were originally neutral with respect of their desirability. They were, simply, places that didn't exist. Hence George Orwell described his 1949 novel *Nineteen Eighty Four* as a utopia, a "not place", even if it was an unfavourable "not place" to be. Only much later did utopias take on their pejorative association as "places where one *wanted* to be". Literary dystopias only become commonplace in the 1960s.

Dystopias

As noted in Chapter 7, literary scholars have identified declensionism as the dominant trope for speculative climate fiction; in other words, things are going to get worse. It fits a well-developed pattern of the dystopic in environmental writing. When applied to climate change yields one of the most salient and powerful of climate imaginaries: The "apocalyptic". Drawing upon the metaphoric world of the Judeo-Christian apocalypse, this imaginary is rooted in belief about ecological collapse caused by anthropogenic climate change and societal breakdown. It creates a sense of urgency to mobilize for decisive and immediate political action and feeds the rhetoric of declensionist social movements such as Extinction Rebellion (see Box 5.3). Climate dystopias, like those created in Margaret Atwood's Maddaddam trilogy (2003, 2009, 2013), frequently depict dysfunctional and chaotic future worlds that parallel their chaotic climates. It is with good reason that literature professor Michael Boyden writes that "climate change fictions overwhelm us with anticipatory melancholia about the looming end of humanity" (2020: 91).

But it is not just climate fiction. The apocalyptic imaginary also finds expression in many non-fiction forays into the climate future. One out of many examples of this trope is Roy Scranton's *Learning to Die in the Anthropocene: Reflections on the End of a Civilisation* (2019). Scranton depicts a world collapsing in the face of climate change:

> Rising seas, spiking temperatures, and extreme weather imperil global infrastructure, crops, and water supplies. Conflict, famine, plagues, and riots menace from every quarter. From war-stricken Baghdad to the melting Arctic, human-caused climate change poses a danger not only to political and economic stability, but to civilization itself . . . and to what it means to be human.

Scranton's essay offers catharsis from facing this dystopic future – his "learning to die" – but his imaginary is apocalyptic in its contours.

Climate dystopias also frequently inhabit the world of gaming, where they offer theatrical and climactic backdrops against which to secure "the quest". For example, the video game *Frostpunk*, originating from Poland, is set in a dystopian world in which a volcanic event has triggered a colossal global ice age (Obermeister & Honybun-Arnolda, 2019). The game's primary scenario consists of surviving the winter – which gets incrementally colder as time progresses – in "New London": A settlement of survivors clustered around a large coal-powered generator. The player must choose between a number of difficult policies and options to ensure the survival of the population.

Utopias

Starting with Thomas More's *Utopia* from the sixteenth century, utopian imaginaries are not necessarily meant to be realisable. But they are intended to provoke aspirational projects and to provide moral and ethical direction for their realisation. The utopian climate imaginary must suspend belief in the gloomiest predictions of climate models and offer worlds where people have responded to climate change in timely and resourceful ways. They must bring to life futures where ways of living have adapted to a changing climate and thrived in the process. But utopias seem a hard sell in today's world of climate predictions emanating from global models. In 2015, the geographer Bruce Braun wrote an editorial introduction for a special collection of essays from leading geographers commissioned under the title "Imagining Socioecological Transformation". These were published in the professional journal of the American Association of Geographers. Braun lamented that geographers had not been successful either at imagining or at bringing about alternative futures that challenged existing injustices and ecological decline. He located the reasons for this in geography's "inability to imagine alternatives in its widespread . . . rejection of prescriptive approaches to political or ecological change and its *widely held suspicion of utopian thought*" (Braun, 2015: 239, emphasis added).

It is not just geographers who may be suspicious of utopias. It is also – as shown in Chapter 7 – the rarity of finding writers of speculative climate fiction who can bring climate utopias to life through their writings. But there are examples. One such comes from the suggestion of the Swedish literary scholar Paul Tenngart who promotes a literary genre that offers positive conceptual infrastructures. He labels this the "climate romance" (Benner et al., 2019), with its emphasis on "self-identification, norm-setting, clear projections, as well as a positive outlook ('happy ending') on the future" (p. 22). Kim Stanley Robinson's novel *New York 2140* is offered by Tenngart as an example of such a utopian climate imaginary, with its positive outlook on the ability of New Yorkers to develop adaptive capacities and sustainable livelihoods even as the future climate changes.

The social imaginaries mobilised by the transformative radicalism examined in Chapter 5 should also be included in the category of climate utopias. This vision of future social transformation as a response to climate change has gained salience over the last decade or more. It is part of the realisation of the profundity of change that climate change might (need to) bring about. A good example of such a utopic (transformative) climate imaginary[3] can be found in the work of two young German scholars, Madita Standke-Erdmann and Alina Viehoff. In their article "The World in 2040: A World Transformed by Climate Justice", published in 2020, they offer the following speculative vision of a future climate utopia:

> Imagine this. It's Monday, 20 February 2040.

The world has changed significantly. About 35 percent of the world's economies have gone almost carbon neutral. We are approaching the 10th anniversary of the Global Climate Justice Agreement of 2030. Representatives from all regions of the world will gather in Bogotá in April to renew this agreement for another 10 years. It is on the brink of becoming the most important and effective legally binding international document in the history of the Anthropocene. Once again, the planet's future lies in the hands of those representatives who are about to pool forces in Colombia's capital. While we can chalk up a victory of having achieved an average increase in temperatures of just below 2°C above pre-industrial levels, life on Planet Earth remains constrained. Hopes are high that the developed "Embracing Earth" principle, which puts the protection of climate and the environment first, will continue to shape the fight for climate justice as a global enterprise. More specifically, securing climate justice rather than securitizing climate change and asking who is affected most has provided the global community with a paradigm shift. So, what changed and where could it lead us? (p. 348)

Technoclimates

The contrast between utopic and dystopic climate futures is well illustrated by the prospect of deploying geoengineering technologies, emerging from what more generally might be described as the imaginary of technoclimates. Some of these technologies – notably solar geoengineering – were considered earlier in the book in relation to the reformed modernist position (Chapter 3) and with respect of responsible research and innovation governance (Chapter 9). Here, I consider them in the context of the sociotechnical imaginaries that inspire and legitimate them. The desire to master and materially redesign the world according to human need or want – and the belief that it is possible to do so – is longstanding. This desire has worked powerfully in the case of climate, in times past offering utopic visions of "perfectable climates" (Fleming, 2010). Many of the ideologies, beliefs, values, and instincts that agitate the politics of climate change are nowhere more clearly on display than in the vigorous arguments about the desirability of technologies to deliberately intervene in the physical functioning the climate system.

This technoclimate imaginary can embrace a variety of geoengineering technologies that seek either to cool the planet or to protect human societies from extreme future climate scenarios. All of them, however, should be considered speculative. In this sense these imagined geo-technologies are another speculative futuring practice. They include solar geoengineering technologies as discussed earlier, with their associated metaphors of "repairing", "restoring", "caring", or "engineering" the Earth's climate. But there are many other proposed schemes: Technologies for artificially regrowing Arctic sea ice and mountain glaciers; massive walls built on the ocean floor to arrest the slide of Antarctic ice sheets into the Southern Ocean; dams across the North Sea to protect the coastlines of northern Europe from a rising ocean; and, most bizarrely, spraying salt across the Arctic Ocean to deliberately keep it ice-free, thereby cooling the world's oceans (Hunt et al., 2020).

The sociotechnical imaginary of technoclimates provokes strong reactions. Techno-optimists see great possibilities in using the ingenuity of engineering technologies to manipulate the Earth's climate in ways to bring it under direct human control. These again usually think in global terms: They are "displaced" from geographical realities. If global emissions of greenhouse gases cannot be reduced or eliminated through conventional means, they would say, then more direct means of offering protection from climate change are warranted. This argument appeals to assorted contrarians – who are already sceptical that emissions reductions are either necessary or feasible – and to some ecomodernists. Other reformed modernists fall on either side of the issue. The argument does not so much revolve about the efficacy of such technologies, although important questions are asked. The dispute is more about responsible innovation (see Chapter 9) and the social, ethical, legal, and political regulation of the technologies. Despite these concerns, some reformed modernists will argue that climate engineering may well be necessary to deploy at some point in the future if planetary temperature is to remain within the warming limit of 2°C embedded in the PACC. At the very least, they say, the technical feasibility and public acceptability of such technologies should be researched (Long, 2020).

Techno-pessimists, on the other hand, see only trouble ahead if these types of climate interventions are pursued. Solar geoengineering may not quite lead to the disastrous outcome of an accidentally induced ice-age, as fictionalised in the 2013 movie *Snowpiercer*. But the actual outcomes are far from assured. Critiques of this form of (global) technological solutionism can be either pragmatic or principled. And they can be voiced amongst the ranks of climate radicals, some reformers, many subalterns (if asked), and some others who, because of their religious cosmology, steer clear of the perceived hubris of the technoclimate imaginary.

Some of these speakers claim pragmatically that the technologies either would not work or could not safely be governed. Others argue on normative grounds that they should not be *allowed* to work. Eileen Crist speaks for these latter when voicing her opposition to the human arrogance of the idea of technoclimates. They reveal, says Crist (2018: 1244), "a Promethean impulse to sustain human hegemony while avoiding the most expeditious approach to the ecological predicament", namely scaling down and pulling back human demands on the natural world. To the extent that their voice has been heard, many subalterns in the Global South are extremely sceptical about solar geoengineering, expecting that it would entrench existing power relations. Some subaltern associations, such as Hands Off Mother Earth (HOME), compare the prospect of Northern-guided technoclimates to the histories of empire. They are once again "taking control of our rain". HOME advocates for a pre-emptive moratorium on geoengineering research.

Geographers can bring important questions to bear on these debates, inserting considerations about space, scale, justice, power, and geopolitics. For example, Bellamy and Palmer (2019) argue against an early foreclosure of *all* such options on the grounds that the spatial scale and efficacy of such technologies is not predetermined, that different governance arrangements can be negotiated, and that questions about their implications for social justice need proper scrutiny rather than premature condemnation. It is certainly the case that some have argued that due attention to climate justice warrants such research and prospective deployment (Horton & Keith, 2016). Others, however, argue just the opposite (McLaren, 2018).

Climate change forever

The idea of climate change is powerful. And it is powerfully creative. When applied to the future, it offers fertile soil in which different social imaginaries take root, expressing a wide range of human hopes, anxieties, and fears. This chapter has explored some of these climate futures using the idea of the climate imaginary. Climate imaginaries are created through social processes that meld together stories, ideologies, values, institutions, and technologies. They shape the practices, lived experiences, and identities of different social groups and provide a collective orientation towards the climatic future. Some imaginaries instinctively default to the global; others may be grounded in specific places. Geography matters. And there are many different futuring practices that underpin climate imaginaries, a few of which have been surveyed here. Some imaginaries gain strength from realist practices, others from more speculative ones; some signal dystopic futures, others more hopeful utopias.

The different climate imaginaries that futuring practices put into public circulation are not merely creative fancies. How future climate is imagined conditions how people respond to the idea of climate change. Climate imaginaries are therefore also political statements. They embody – and make visible – scientific predictions, scenarios, images, metaphors, speculative fiction, and speculative technologies in ways that catalyse political arguments around the goals of public policy. Competing climate imaginaries provide the grit of political contention in the present and they fuel political, moral, and ethical instincts with regard to the future. "Futures are always a normative proposition", explains Ted Nordhaus (2019: para. 32), writing with respect to climate change and eco-identities, "and the increasing fragmentation and enfranchisement of social and cultural identities have made it ever more difficult to sustain social consensus around any explicit, shared vision of a good life or a desirable future". Contrary to those who search for a "unifying strategic narrative" about climate change around which all political actors may gather (Bushell et al., 2017), climate imaginaries offer exactly the opposite. This book has shown that the idea of climate change gathers multiple meanings as it travels around the world and it engenders many different political projects. The diversity of climate imaginaries – the different scales they project onto and their often antagonistic relationship with each other – demonstrates how implausible it is to reduce climate change to a "single strategic narrative".

On the other hand, it *is* clear that earthly climates will continue to change, notwithstanding human efforts to arrest or redirect this change. Restoring global climate to some past condition is not an option, even if local climates may yield to modifying technologies. Neither is it possible to retrieve some *imagined* condition of past climate stasis. As I explained in Chapter 1, this imaginative condition was the comforting assumption of a place's settled climate, an assumption that provided the imaginative resources needed to enact a secure livelihood and enable purposeful action in the world. An indefinite future of a physically changing climate, now brought about largely by human hands, has to be confronted. But also to be grasped is the fact that the *idea* of an unsettled climate is with us forever.

Human imaginations will continue to create an unquenchable assortment of possible futures. But as this chapter has shown, many of these futures are now deeply conditioned on how climate is imagined to turn out. As shown in Chapter 1, historical and cultural evidence suggests that a belief that a world falling apart is the cause of climate change may be just as powerful as the belief that a change in climate will unsettle the world. Climate imaginaries therefore offer not only descriptions of possible future disturbances to physical processes that have world-changing effects. They also capture in climatic terms the possible consequences of what is perceived to be the unsettling of people's present and future social worlds.

So, this is how things stand in 2021. Climate change is not just an unavoidable physical reality, its consequences laying bare the injustices of yesterday's, today's, and tomorrow's worlds. But the *idea* of climate change is also now inescapable. It is firmly embedded in cultural imaginations. But *as* an idea, it at least offers us a new language and fresh imaginative resources with which to pass judgement on the world and to prepare for what may lie ahead.

And what may lie ahead? In conclusion, I offer a collection of imaginative vignettes of the climate futures that await us. They speak as loudly as do the scientific assessments of the IPCC.

It is "the future".[4] A Barefoot Hopi, from the Native American Pueblo tribe, is a prime mover of the international convention called by natural and Indigenous healers to discuss the Earth's crisis. A main concern is the changing weather. At the convention, the Barefoot Hopi speaks eloquently, warning of the Earth's "outrage": "The rain clouds no longer gather; the sun burns the Earth until the plants and animals disappear and die". Delivering the conference keynote, the Hopi speaks of the international coordinated effort in which he plays a central part, traveling to Africa, Asia, and around the world to meet with Indigenous tribal people. The transnational coalition that results emphasises similarities between the tribal people of Africa and the tribal people of the Americas, especially in terms of settler colonial land loss to whites. The Hopi accuses the United States of eco-terrorism, poisoning the water and air.

He predicts earthquakes, tidal waves, landslides, drought, and wildfire before it's all over but also promises that "a force [is] gathering," coming from the South, to "counter the destruction of Earth".

It is the year 2031.[5] The attempt back in 2014 to stop climate change through solar geoengineering catastrophically backfired, creating a new ice age. The remnants of humanity took to a circumnavigational train – the thousand-carriage Snowpiercer – operated by a reclusive transportation magnate. Now, 17 years later, in an ice-cold world, the passengers on the train have become segregated. The elite reside in the extravagant front carriages, while the poor are confined to the squalid tail compartments, controlled by armed guards. As the ice begins to thaw and climate warms, the tail passengers revolt leading to an explosion that derails the train. As the two child protagonists escape the wreckage, they see a polar bear in the distance, indicating that life still exists outside the train.

It is the year 2040.[6] Rising temperatures, extreme weather, and the rise in sea levels have had severe consequences for ecosystems, affecting not only flora and fauna but also human populations. However, there is also good news. Countries like Costa Rica and Ecuador have gone 90 per cent carbon neutral. These countries' environmental protection policies are holistic, following the principle of "Embracing Earth", which has also started to influence large economies. In Germany, large reforestation endeavours allow the country to be more resilient towards heatwaves, as well as encouraging native species to repopulate areas and to create cooler microclimates. And in China, massive cuts in carbon dioxide emissions have proved to be a major turning point in international climate change governance. A working group of the "Pachamama Earth Trust", the largest and most influential subsidiary of the "Sir David Attenborough Foundation for Global Environmental Protection", is travelling the world in hydroplanes and natural gas–powered ships. They are gathering knowledge about the effective implementation of the goals set in the 2030 Global Climate Justice Agreement.

It is the year 2055.[7] The world is ravaged by catastrophic climate change. London is flooded, Sydney is burning, Las Vegas has been swallowed up by desert, the Amazon rainforest has burnt up, and snow has vanished from the Alps. An unnamed archivist is entrusted with the safekeeping of humanity's surviving store of art and knowledge. Alone in his vast repository off the coast of the largely ice-free Arctic, the archivist reviews archival footage from 50 years ago, trying to discern where it all went wrong. Amid news reports of the gathering effects of climate change and global civilisation teetering towards destruction, he alights on six stories of individuals whose lives in the early years of the twenty-first century illustrate the tragedy of the impending catastrophe.

It is the year 2140.[8] Manhattan has been flooded by a 16-metre rise in ocean level. Although New York is mostly underwater, people still live in the upper floors of the city's buildings. Most of Manhattan below 46th Street is flooded and has earned the nickname "SuperVenice". The wealthy live in newly constructed skyscrapers in Uptown Manhattan and near The Cloisters, as both locations remain above water. These towers are fitted with flood-prevention mechanisms and boat storage. Denver has replaced New York as the centre of American finance and culture and much of the United States has been deliberately abandoned by humans in order to make room for wildlife.

It is the year 2393.[9] The world is almost unrecognizable. Clear warnings of climate catastrophe went ignored for decades, leading to soaring temperatures, rising sea level, widespread drought and the disaster now known as the Great Collapse of 2093. In this year, the disintegration of the West Antarctica Ice Sheet led to mass migration and a complete reshuffling of the global order. An elderly Chinese historian, writing in the era of the Second People's Republic of China and on the 300th anniversary of the Great Collapse, is now recounting and explaining the climatic events of the twenty-first century. These events culminated in what became known as the "second Dark Age", prompted by the Great Collapse. The historian presents a disturbing account of how the political and economic elites of the so-called advanced industrial societies failed to change course in the 2020s and so brought about the collapse of Western civilization.

Notes

1 This is so-named after the Boston Consulting Group who introduced this scenario device into marketing strategy in the early 1970s.

2 Note the obvious difference between precision and accuracy. One can be precisely wrong but also imprecisely correct.

3 Another example of a utopic imaginary is *A Message From the Future With Alexandria Ocasio-Cortez*, available on YouTube at www.youtube.com/watch?v=d9uTH0iprVQ [accessed 26 August 2020].

4 Source: Streeby (2018: 57–59), paraphrased from her commentary on Leslie Silko's 1992 novel *The Almanac of the Dead*. Silko is an American writer, a Laguna Pueblo Indian woman, and her speculative fiction offers powerful Indigenous futurisms of climate change that challenge modernity's narrative of progress.

5 *Snowpiercer* movie, released in 2013 in South Korea, directed by Bong Joon-Ho.

6 Source: Paraphrased from Standke-Erdmann and Viehoff (2020).

7 *The Age of Stupid* movie, released in 2009, directed by Fanny Armstrong.

8 Source: Robinson, K.S. (2017) *New York 2140*.

9 Source: Oreskes and Conway (2014) *The Collapse of Western Civilisation*.

Further Reading

For a general introduction to the field of futurology, **Peter Bowler's** *A History of the Future: Prophets of Progress from H G Wells to Isaac Asimov* (Cambridge University Press,

2017) is to be recommended. Bowler examines the range of futuring practices developed during the twentieth century and shows how the future enters into and resides in the social imagination. To explore the idea of social imaginaries in general, **Benedict Anderson's** *Imagined Communities: Reflections on the Origins and Spread of Nationalism* (Verso, 1983) is the best place to start. In *Scientists as Prophets: A Rhetorical Genealogy* (Oxford University Press, 2013), the rhetorician **Lynda Walsh** explains how scientists, when making predictions about the future, fulfil a prophetic ethos through seeking to persuade their public audiences to take action. For a contrasting approach for thinking about climate futuring, which includes Indigenous futuring practices, read **Shelley Streeby's** *Imagining the Future of Climate Change: World-Making Through Science Fiction and Activism* (University of California Press, 2018). In *Culture and Climate Change: Scenarios* (Shed Press, 2019), **Renate Tyszczuk, Joe Smith,** and **Robin Butler** have brought together a great collection of short essays from leading thinkers reflecting on the nature of scenarios of the climatic future. The book is available on-line at: www.cultureandclimatechange.co.uk/site/assets/files/1028/ccc_scenarios_compiled.pdf. The best book for understanding the role and dangers of metaphors in environmental science communication is **Brendon Larson's** *Metaphors for Environmental Sustainability: Redefining Our Relationship with Nature* (Yale University Press, 2011). There are a rapidly growing number of books about technoclimates and the future of climate engineering. **Clive Hamilton's** *Earthmasters: The Dawn of the Age of Climate Engineering* (Yale University Press, 2013) is an informed and well-written warning about the dangers of climate engineering. For a more positive exploration of the implications of these technologies, a good suggestion is **Oliver Morton's** *The Planet Remade: How Geoengineering Could Change the World* (Princeton University Press, 2016). Very helpful for putting future climates into the broader context of human development and technological change is **Christopher Preston's** *The Synthetic Age: Outdesigning Evolution, Resurrecting Species and Reengineering our World* (MIT Press, 2018). And for an excellent account of some of the ways in which different social formations make sense of climate change, read **Candis Callison's** *How Climate Change Comes to Matter: The Communal Life of Facts* (Duke University Press, 2014).

QUESTIONS FOR CLASS DISCUSSION OR ASSESSMENT

Q10.1: Personally, do you find speculative fiction (e.g. Cli-Fi novels) or climate model predictions (e.g. IPCC reports) more persuasive for engaging with the climatic future? Why is this?

Q10.2: What evaluative criteria are authors using when they talk about the possibility of "good" or "bad" Anthropocenes?

Q10.3: What is the allure of technofixes for resolving the problems of climate change? Should they be dismissed out of hand or embraced cautiously or even enthusiastically?

Q10.4: In calls to attend more vigorously to the demands of climate justice, how central are claims about climate's *future* trajectory as opposed to claims about climate's *present* reality?

Q10.5: In your imagination, travel 30 years into the future. How do you think climate change will be being talked about in 2051? Will the idea of climate change still be a future-centred discourse? Imagine what climate change will mean for *you*.

EPILOGUE

The future of an idea

This book has explained the idea of climate change in the past, present, and future. It is a task I have approached using the training and instincts of a geographer. In this brief Epilogue I offer some final reflections on the future of the idea of climate change. How might people in, say, the year 2067 be thinking about climate change? What might the idea of climate change signify for them in terms of daily living, creative practice, religious belief, technological innovation, and political action? This is very different question than asking climate scientists "What will the global temperature or the level of the ocean be in 2067?" But first, a short reprise of the book.

Chapter 1 traced the origins, ubiquity, and functions of the idea of climate change in the cultural imaginations of premodern and non-scientific worlds. The idea of climate change only took on its distinctive modernist form in the late twentieth century, as explained in Chapter 2. But this is only the latest stage in an evolving story about how people think about their relationship with climate. Indeed, in describing three "science-based" positions on climate change, Chapters 3, 4, and 5 advanced the argument that the idea of climate change – and especially the science of climate change – becomes enrolled in a struggle between competing visions of how societies should be organised, of human well-being and emancipation. (As Chapter 10 has shown, this struggle is about control of the future.) Using the idea of "modernity" as a heuristic, I argued at the end of Chapter 5 that those voices labelled *sceptical contrarians* may be understood as affirming "Modernity 1.0", *reformed modernisers* as proposing "Modernity 2.0", and *transformative radicals* as being "Anti-Modern" or perhaps "Beyond-Modern". The idea of climate can, it would seem, provoke either a different modernity, more modernity, or less modernity.

Whilst these political struggles for the climatic future are frequently couched in terms of "what the science says", I have also made clear that there are other equally powerful, perhaps even more powerful, ways of making sense of the idea of climate change. These have been presented in the three "more-than-science" approaches explored in Chapters 6, 7, and 8. To know how to make sense of climate change or how to respond to it, it is not enough to "follow the science" or simply "listen to scientists". To the contrary, by listening to subaltern voices, by immersing oneself in the creative worlds of artists, or by sharing in the life of religious communities, the limits of climate science for sense-making about climate change will be revealed. Although writing more generally about the limits of science, political theorist Erin Dolgoy's observation applies perfectly to the case of climate change: "Science cannot account for how the individual experiences things and therefore scientific explanations cannot capture the very essence of the experience" (Dolgoy, 2016: 247). To discover what the experience of climate change means requires moving beyond science.

So, what about the future of the *idea* of climate change? It is one thing to develop an intellectual history of an idea. And it is possible – as this book has also done – to interrogate the multiple meanings of climate change in today's world. But how might the idea of climate change itself change over the next 25, 50, or 100 years? How does one write about the future of an idea? It is hard to explain the future of an idea when the very notion of "the future" itself has a past and a future (Bowler, 2017). Yet there seems no shortage of ideas for which their future has been written – for example, the ideas of progress, democracy, science, socialism, capitalism, intelligence, being human. The list is long.

So where to start? I can state with some confidence that in 2067 physical climates around the world will be warmer than they are now, there will be less ice on the planet, and the ocean level will be higher. The climates of 2067 will be different in many other respects as well. Of this much science can be certain. More tentatively, a range of environmental changes that may follow from this warming – and some possible social impacts – can be sketched. This after all is what the IPCC assessments are designed to deliver, building on the findings of scientific and social scientific research. But this is not the same as predicting how people in 2067 will be *thinking* about climate change. The research assessed by the IPCC doesn't help us understand how the *idea* of climate change will evolve.

We encountered this problem in reverse in Chapter 1 when trying to imagine how William Shakespeare in Elizabethan England or Dutch maritime traders in the seventeenth century thought about the possibility of a changing climate. Perhaps a little easier is to reconstruct how climate change was understood in different cultures in the 1970s or the 1920s. This after all is within the lifetime of our parents or grandparents. The physical climate of these past decades was different to today. This is what historical climatology can reconstruct for us. But also different was the way that people thought about climate change in these earlier decades. To assist with *this* reconstruction, we have extant documentary material to rely on, as well as oral histories and living memories.

But to describe the future of the idea of climate change presents a different and still harder task. Nevertheless, let me suggest three broad framings of the idea of climate change that might help us imagine some of its possible future trajectories.

Climate change as an engineering problem

In the late decades of the twentieth century, climate change – understood as an emergent phenomenon of modernity – was initially framed as an engineering problem. This framing was prompted by intensified scientific monitoring of the planet and by the development of numerical simulation models of how the Earth System functioned. Chapter 2 tells this story. The methods of science and the tools of engineering could be used, first, to diagnose and then, hopefully, to remedy the increasing disturbance to climate being caused by humanity's deepening dependence on fossil fuels and by its rapid transformation of the surface of the planet. The climate system was looked upon much like a malfunctioning car or a diseased body. Much of this thinking survives today. Climate change is reduced to an engineering challenge of how to deliver a world only 2°C warmer than the late nineteenth century, of how to manage the carbon budget to net-zero emissions by 2050 or some such date. As we have seen in earlier chapters, the engineering involved in reaching these goals may be technological, economic, or behavioural. But *engineering problems* demand solutions. If the proposed control technologies won't work, then more radical engineering solutions need pursuing. Inspired by the imaginary of technoclimates – as discussed in Chapter 10 – re-engineered climates can be imagined, designed, simulated, and, maybe, finally delivered.

This line of thinking may have a future. For example, a newly opened centre at the University of Cambridge – the Cambridge Centre for Climate Repair – has embarked on the task of researching solutions to repair a "damaged climate", to return critical elements of the climate system to an undamaged state. In *AI and the Environmental Crisis: Can Technology Really Save the World?* (2020), Keith Skene – Director of the virtual Biospheric Research Institute – offers a glimpse into one such possible world. Skene argues that for artificial intelligence (AI) to deliver on its promise of a better world, it needs its foundations rooted in what he calls "ecological intelligence" – an appreciation of the interconnectedness of all things – rather than in human intelligence. Since human existence is dependent upon the Earth System, so must the goals of artificial intelligence be set within this context. Skene exposits on how ecological intelligence might be employed as a tool to deliver – through the agency of AI systems – an engineering solution to the problem of climate change. His faith is that AI can usher in the greatest transformation in human history.

Climate change as the locus of politics

An alternative framing of the idea of climate change is to see it as a new and increasingly dominant locus for human politics. In this line of thought, climate change is neither a problem to be solved nor a process to be stopped. At least not directly. Rather, climate change is a malleable idea, around which gathers different ideologies, values, interests, and imaginaries that form the grit of political life. Contrary to what is claimed by some (see Box 5.2), climate change needs to become *more* political rather than *less*. The idea of climate change releases a powerful new dynamic into public life. It re-animates the age-old political struggle about how societies should be organised, what constitutes a well-nourished and meaningful human life, and how these goals can be achieved.

The reason the idea of climate change possesses such political agency is that it provokes new ways of connecting human and material worlds and of thinking about their deepening interdependencies. This provocation of climate change to modern thought has been a gradual process. For example, the new Earth System Science paradigm, designed by NASA in 1986, connected in systemic fashion the multi-scalar physical and ecological dynamics of the planet across all scales, from micro to global. The Brundtland Report of 1987 (WCED, 1987) brought together the domains of economy, society, and environment. It advocated a more holistic way of thinking about future human development, a proposal for which the idea of climate change has given sustained impetus. And in recent years the idea of climate change has provoked new creative thinking about the interplay between the forms of knowledge offered by the sciences, Indigenous communities, creative arts, religions, and other lifeforms. Chapters 6, 7, and 8 tell this latter story.

Engineering problems need solutions, but *politics* requires contestation and struggle. The successful animation of political life depends upon societies possessing institutions and processes that allow different interests to be heard, debated, and accommodated in public forums. The idea of climate change challenges this ambition. Climate change is an idea of such size, scope, and imaginative power that it escapes the capacity of any one person to grasp and for political institutions to resolve. It sets in motion new and unfamiliar ecological pressures, social dynamics, ethical quandaries, and political complexities. Many of chapters in Section 2 of the book elaborated these challenges. They strain the ability of the political processes of national jurisdictions to accommodate (Chapter 9). These challenges increase the possibility of "failed states", political entities that disintegrate to a point where the basic conditions and responsibilities of a sovereign government no longer function properly. Failed states precipitated by climate change may turn out to be more dangerous for the world that failing ice sheets.

And these challenges spill over beyond the confines of one nation, ethnicity, religion, culture, or history. In this way of thinking about climate change, all politics become climate politics, to a greater or lesser degree (Kelly, 2019). Dealing with this new demanding world of climate politics will require careful and constant attention to the nature and condition of the political institutions and processes that different jurisdictions have available to them. The idea of climate change will provoke new forms of earthly politics, beyond the old categories of Left and Right, collectivist or libertarian. "Learning new ways to inhabit the Earth is our biggest challenge", observes Bruno Latour in *Down to Earth: Politics in the New Climatic Regime* (2018). "Bringing us down to earth is the task of politics today".

Climate change as human predicament

A third framing of the idea of climate change through which to think about its future trajectory is as a human predicament. Climate change presents humanity – all of "us" – but in starkly different ways, with an unpleasant and confusing situation that is impossible to solve or eradicate. The predicament arises partly from the legacy of the past, a past that none of us chose but that we all inherit. This is the fossil-fuelled industrial past handed down to the world by the more developed nations. And it is the colonised and oppressed past experienced by subalterns in less developed nations. These legacies are inscribed in today's atmosphere and embedded in the climatic vulnerabilities of exposed communities. But the predicament also arises from what the present generation is, in effect, borrowing blindly from the future. People's lives are wedded to hard infrastructures and soft institutions, to social practices, aspirations, and value systems that are not easy to change. The future is being mortgaged, not sure the mortgage can ever be repaid. To sustain eight billion people under this model, with at least two billion more people to come, seemingly leaves little room for manoeuvre. The technical term is "lock-in".

Climate change as predicament thus becomes a magnifying mirror. Looked into, we see more clearly what is at stake with the choices that remain with us. The magnifying mirror of climate change reflects our human condition, yet deepens its contradictions. It reflects our agency in the world, whilst revealing its limitations. It reflects our fears for the future but also amplifies them. And it reflects our instincts for justice and fairness, whilst frustrating our realisation of them.

Engineering problems require solutions and *politics* needs agonistic spaces to enact and resolve struggles. Problems call for solutions; the question becomes whether solutions can be found and made to work. But climate change is a wicked problem, to which no solution is possible. So, climate change becomes the new locus of politics, manifesting itself in constant unresolved struggle. But a predicament, by contrast, can neither be solved through engineering nor resolved through politics. A predicament just won't go away. What *predicaments* need are stories. Interpretative stories – what some may call guiding myths – through which to understand the predicament and to come to terms with it.

Some of these narrative resources may be found in non-Western cultures and in religious traditions, as suggested in Chapters 6 and 8. Some may need bringing to life through the creative instincts of writers, artists, and poets, as witnessed in Chapter 7. Tragedies, comedies, romances, laments, tricksters, and prophets may each, in their own ways, help people recognise and accept the predicament in ways that science alone cannot. To live with it – but also to move on. Framing climate change as a predicament is not to retreat from public life. Nor is it to abandon the politics of hope. It is to engender a pragmatic and humbler stance with respect to innovation, technology, politics, justice, and the future. Accepting climate change as a predicament also makes it easier to accept the limits of human prescience, agency, and integrity. Paradoxically, this allows hope to be sustained. It allows the possibility of solving some, more modest, problems, without needing to solve the "unsolvable" problem of climate change.

The American Protestant theologian Reinhold Niebuhr (1892–1971) spent his life seeking to relate the Christian faith to the realities of politics and diplomacy. His oft-repeated prayer – "God, grant me the serenity to accept the things I cannot change, the courage to change the things I can, and the wisdom to know the difference" – captures the combination of realism and activism that I believe framing climate change as predicament offers.

The idea of climate change in geography

So, is the idea of climate change in the future most likely to be characterised as an engineering problem, as a new locus of politics, or as a human predicament? Or as a combination of all three? Will the idea evolve in ways yet hard to see? The idea of climate change in today's world certainly attracts adherents to all three of these framings. And they are ones that geographers can not only relate to – but help elucidate – as this book has sought to do. The discipline of geography embraces the natural sciences, social sciences, and humanities, conventional traditions of intellectual inquiry that broadly map onto the three framings offered previously.

My own view is that I am doubtful that framing climate change as engineering problem will take us very far. I believe it sets up the world for inevitable – and dangerous – failure. It is already clear that the limit of 2°C of warming will likely be exceeded. And yet continuing to demand that this limit not be exceeded will lead to ever more draconian or far-fetched engineering solutions being proposed and implemented. The problem with the engineering analogy applied to the climate system is obvious. Even if the physical processes of the planetary climate are amenable to engineering interventions, which is questionable, there happen to be eight billion people and 197 national entities, with their unruly social and cultural attachments and political ambitions, whose behaviour cannot be reduced to a numerical calculus. The Earth's climate is not like an engineered dam, an autonomous vehicle, or a nuclear reactor. Even in these cases, the limits of control technologies are clearly seen. How much more do such limits apply in the case of engineering the entirety of sentient life and inanimate matter on the planet?

More productive, I believe, is the interplay between framing climate change as a predicament and as the new locus of politics. There remains a tension between these two positions; the inability to find political solutions to climate change is part of the predicament. But as I have suggested in earlier places in this book, it is possible to use the idea of climate change creatively to bring about desirable change in the world without remaining hostage to the impossible dream of subjecting the condition of global climate to human will.

Whatever your personal views may be, I leave you with the claim that geographers are well-placed to identify and articulate multi-faceted interpretations of the idea of climate change. Geographers are trained to recognise the importance of space, place, scale, and difference in social and ecological dynamics and in their interactions. They recognise the mutual shaping of natural and cultural worlds out of which emanates the phenomenon of climate change. And geographers understand that human predicaments are never solely material, technical, or ecological and they therefore embrace enthusiastically the political, the ethical, and the spiritual.

In an essay titled "Navigating climate's human geographies: exploring the whereabouts of climate politics", geographer Harriet Bulkeley reflected on what moving from climate change-as-problem to climate change-as-condition (what I call predicament) might mean for the discipline of geography. How might geographers, she asks, allow the idea of climate change to shape their professional practice? Geographers cannot evade the idea of climate change in their efforts to "graph-the-geo", to write and draw the world. After all, climate change is redrawing the material and imaginative worlds that geographers study. But this realisation is not, she says, "a call for us all to become climate change . . . geographers, but instead to recognize that work in our discipline may be changed by climate change" (Bulkeley, 2019: 40).

The first necessity of the recovering addict is to own their behaviour, to recognise what they have become, to face who they are. By analogy, these are lessons to learn with respect to climate change. Unintended and uninvited, climate change has accompanied the pasts, both the savoury and the unsavoury, whose legacies are now inherited – the imperial, industrial, economic, political, military, and cultural pasts. It is what "we" humans have done to the world. Yes, our culpabilities are differentiated and our capacities for change are unequal. But together we must accept, own, and then face a predicament that is not of our choosing.

Will climate change ever be stopped? No, not decisively. Will the idea of climate change mutate? Constantly, in ways that will continue to surprise us.

BIBLIOGRAPHY

Abbott, K.W. (2012) The transnational regime complex for climate change. *Environment and Planning C: Government and Policy*. 30: 571–590.

Agarwal, A. and Narain, S. (1991) *Global Warming in an Unequal World*. Delhi, India: Centre for Science and the Environment.

Allan, B.B. (2017) Second only to nuclear war: Science and the making of existential threat in global climate governance. *International Studies Quarterly*. 61: 809–820.

Amri, U. (2014) From theology to a praxis of "eco-jihad": The role of religious civil society organisations in combating climate change in Indonesia. (pp. 75–93) In: *How the World's Religions are Responding to Climate Change: Social Scientific Investigations*. Veldman, R., Szasz, A. and Haluza-Delay, R. (eds.). Abingdon: Routledge.

Anderson, K. and Bows-Larkin, A. (2013) Avoiding dangerous climate change demands de-growth strategies from wealthier nations. *kevinanderson.info*. Available at https://kevinanderson.info/blog/avoiding-dangerous-climate-change-demands-de-growth-strategies-from-wealthier-nations/ [Accessed 12 August 2020].

Andonova, L. (2020) Does successful emissions reductions lie in the hands of non-state rather than state actors? YES: Because it requires commitments by all actors, private and public. Chapter 12 (pp. 177–182) In: *Contemporary Climate Change Debates: A Student Primer*. Hulme, M. (ed.). Abingdon: Routledge.

ARC (Alliance of Religions and Conservation) (2007) UN and ARC launch programme with faiths on climate change. 7 December. Available at www.arcworld.org/news.asp?pageID=207 [Accessed 25 June 2020].

Armstrong, K. (2005) *A Short History of Myth*. Edinburgh: Canongate.

Asafu-Adjaye, J., Blumqvist, L., Brand, S. and co-authors (2015) *An Ecomodernist Manifesto*. Available at www.ecomodernism.org/manifesto-english [Accessed 14 August 2020].

Atwood, M. (2003) *Oryx and Crake*. London: Bloomsbury.

Atwood, M. (2009) *The Year of the Flood*. London: Bloomsbury.

Atwood, M. (2013) *MaddAddam*. London: Bloomsbury.

Aykut, S.C. (2016) Taking a wider view on climate governance: Moving beyond the 'iceberg,' the 'elephant,' and the 'forest'. *WIREs Climate Change*. 7(3): 318–328.

Aykut, S.C., Morena, E. and Foyer, J. (2020) 'Incantatory' governance: Global climate politics' performative turn and its wider significance for global politics. *International Politics*. https://doi.org/10.1057/s41311-020-00250-8.

Baker, Z. (2021) Agricultural capitalism, climatology and the 'stabilization' of climate in the United States, 1850–1920. *British Journal of Sociology*. 72: 379–396.

Ballard, J.G. (1962) *The Drowned World*. London: Victor Gollanz.

Barnett, L. (2015) The theology of climate change: Sin as agency in the Enlightenment's Anthropocene. *Environmental History*. 20: 217–237.

Barnett, L. (2019) *After the Flood: Imagining the Global Environment in Early Modern Europe*. Baltimore: Johns Hopkins University Press.

Bastin, J-F., Finegold, Y., Garcia, C., Mollicone, D., Rezende, M., Routh, D., Zohner, C.M. and Crowther, T.W. (2019) The global tree restoration potential. *Science*. 365: 76–79.

Bate, J. (2000) *The Song of the Earth*. London: Picador.

Bates, P. (2007) Inuit and scientific philosophies about planning, prediction, and uncertainty. *Arctic Anthropology*. 44(2): 87–100.

Beck, S. and Mahony, M. (2018) The IPCC and the new map of science and politics. *WIREs Climate Change*. 9(6): e547.

Beck, U. (1997) Global risk politics. (pp. 18–33) In: *Greening the Millennium? The New Politics of the Environment*. Jacobs, M. (ed.). Oxford: Blackwell.

Beck, U. (2009) Critical theory of world risk society: A cosmopolitan vision. *Constellations*. 16(1): 3–22.

Beeson, M. (2019) *Environmental Populism, the Politics of Survival in the Anthropocene*. London: Palgrave Macmillan.

Bellamy, R. and Osaka, S. (2019) Unnatural climate solutions? *Nature Climate Change*. 10(2): 98–99.

Bellamy, R. and Palmer, J. (2019) Geoengineering and geographers: Rewriting the earth in what image? *Area*. 51(3): 524–531.

Bendell, J. (2018) *Deep Adaptation: A Map for Navigating Climate Tragedy*. IFLAS Occasional Paper 2. University of Cumbria, UK. Available at www.lifeworth.com/deepadaptation.pdf.

Benner, A-K., Rothe, D., Ullström, S. and and Stripple, J. (2019) *Violent Climate Imaginaries: Science-Fiction-Politics*. Hamburg: Institute for Peace Research and Security Policy.

Bernstein, S. and Hoffman, M. (2019) Climate politics, metaphors and the fractal carbon trap. *Nature Climate Change*. 9(12): 919–925.

Berry, E. (2022) *Climate Politics and the Power of Religion*. Bloomington: Indiana University Press.

Bhavnani, K.K., Foran, J., Kurian, P.A. and Munshi, D. (eds.) (2019) *Climate Futures: Reimagining Global Climate Justice*. Chicago, IL: University of Chicago Press.

Biermann, F. (2014) *Earth System Governance: World Politics in the Anthropocene*. Cambridge, MA: MIT Press.

Bobbette, A. (2019) Priests on the shore: Climate change and the Anglican church of Melanesia. *GeoHumanities*. 5(2): 554–569.

Boia, L. (2005) *The Weather in the Imagination*. London: Reaktion Books.

Bottoms, S. (2012) Climate change 'science' on the London stage. *WIREs Climate Change*. 3(4): 339–348.

Bowler, P. (2017) *A History of the Future: Prophets of Progress from H G Wells to Isaac Asimov*. Cambridge: Cambridge University Press.

Boyden, M. (2020) The pathogenesis of the modern climate. *Ecozona*. 11(1): 80–98.

Boyers, R. (2019) *The Tyranny of Virtue: Identity, the Academy, and the Hunt for Political Heresies*. New York: Scribner.

Boykoff, M.T. (2011) *Who Speaks for the Climate? Making Sense of Media Reporting on Climate Change*. Cambridge: Cambridge University Press.

Brace, C. and Geoghegan, H. (2011) Human geographies of climate change: Landscape, temporality, and lay knowledges. *Progress in Human Geography*. 35(3): 284–302.

Brandão, I. (1985) *And Still the Earth. A Novel*. New York: Avon Books.

Brandt Commission (1980) *North South. A Programme for Survival. Report of the Independent Commission on International Development Issues*. London: Pan Books.

Braun, B. (2015) Futures: Imagining socioecological transformation – An introduction. *Annals of the Association of American Geographers*. 105(2): 239–243.

Bravo, M. (2009) Voices from the sea-ice: The reception of climate impact narratives. *Journal of Historical Geography*. 35(2): 256–278.

Bremer, S. and Meisch, S. (2017) Co-production in climate change research: Reviewing different perspectives. *WIREs Climate Change*. 8(6): e482.

Bridge, G., Bulkeley, H., Langley, P. and van Veelen, B. (2020) Pluralizing and problematizing carbon finance. *Progress in Human Geography*. 44(4): 724–742.

Buell, L. (1995) *The Environmental Imagination: Thoreau, Nature Writing, and the Formation of American Culture*. Cambridge, MA: Harvard University Press.

Bulkeley, H. (2016) *Accomplishing Climate Governance*. Cambridge: Cambridge University Press.

Bulkeley, H. (2019) Reflections on navigating climate's human geographies. *Dialogues in Human Geography*. 9(1): 38–42.

Büntgen, U. (2019) Re-thinking the boundaries of dendrochronology. *Dendrochronologia*. 53: 1–4.

Büntgen, U., Myglan, V.S., Charpentier Ljungqvist, F., McCormick, M., Di Cosmo, N. and co-authors (2016) Cooling and societal change during the late antique little ice age from 536 to around 660 AD. *Nature Geoscience*. 9: 231–236.

Bushell, S., Buisson, G.S., Workman, M. and Colley, T. (2017) Strategic narratives in climate change: Towards a unifying narrative to address the action gap on climate change. *Energy Research and Social Science*. 28: 39–49.

Callison, C. (2014) *How Climate Change Comes to Matter: The Communal Life of Facts*. Durham, NC: Duke University Press.

Cameron, E.S. (2012) Securing indigenous politics: A critique of the vulnerability and adaptation approach to the human dimensions of climate change in the Canadian arctic. *Global Environmental Change*. 22(1): 103–114.

Cameron, F., Hodge, B. and Salazar, J.F. (2013) Representing climate change in museum space and places. *WIREs Climate Change*. 4(1): 9–21.

Carbon Brief (2018) How carbon finance flows around the world. *Carbon Brief: Clear on Climate*. 6 December. Available at www.carbonbrief.org/interactive-how-climate-finance-flows-around-the-world [Accessed 13 August 2020].

Carey, M. (2012) Climate and history: A critical review of historical climatology and climate change historiography. *WIREs Climate Change*. 3(3): 233–249.

Carolan, M. (2010) Sociological ambivalence and climate change. *Local Environment*. 15(4): 309–321.

Castree, N. (2013) *Making Sense of Nature*. Abingdon: Routledge.

Chakrabarty, D. (2009) The climate of history: Four theses. *Critical Inquiry*. 35(Winter): 197–222.

Chan, S., Boran, I., van Asselt, H. and co-authors (2019) Promises and risks of nonstate action in climate and sustainability governance. *WIREs Climate Change*. 10(3): e572.

Chaturvedi, S. and Doyle, T. (2015) *Climate Terror: A Critical Geopolitics of Climate Change*. London: Palgrave Macmillan.

Chiari, S. (2019) Climatic issues in early modern England: Shakespeare's views of the sky. *WIREs Climate Change*. 10(4): e578.

Chilvers, J. and Kearnes, M. (eds.) (2015) *Remaking Participation: Science, Environment and Emergent Publics*. Abingdon: Routledge.

Clack, C.T.M., Qvist, S.A., Apt, J. and co-authors (2017) Evaluation of a proposal for reliable low-cost grid power with 100% wind, water, and solar. *Proceedings of the National Academy of Sciences*. 114(26): 6722–6727.

Clingermann, F., O'Brien, K.J. and Ackerman, T.P. (2017) Character and religion in climate engineering. *Issues in Science and Technology*. Fall Issue: 25–28.

Cloud, D. (2020) The corrupted scientist archetype and its implications for climate change communication and public perception of science. *Environmental Communication*. 14(6): 816–829.

Coen, D.R. (2011) Imperial climatographies from Tyrol to Turkestan. *Osiris*. 26: 45–65.

Coen, D.R. (2018) *Climate in Motion: Science, Empire and the Problem of Scale*. Chicago, IL: University of Chicago Press.

Collingridge, D. (1980) *The Social Control of Technology*. London: Pinter.

Cook, J. (2020) Is emphasising consensus in climate science helpful for policymaking? YES: Because closing the consensus gap removes a roadblock to policy progress. Chapter 9 (pp. 127–134) In: *Contemporary Climate Change Debates: A Student Primer*. Hulme, M. (ed.). Abingdon: Routledge.

Corbera, E., Calvet-Mir, L., Hughes, H. and Paterson, M. (2016) Patterns of authorship in the IPCC working group III report. *Nature Climate Change*. 6(1): 94–99.

Coscieme, L., da Silva Hyldmo, H., Fernández-Llamazares, A. and co-authors (2020) Multiple conceptualisations of nature are key to inclusivity and legitimacy in global environmental governance. *Environmental Science and Policy*. 104: 36–42.

Crist, E. (2018) Reimagining the human. *Science*. 362: 1242–1244.

Cruikshank, J. (2001) Glaciers and climate change: Perspectives from oral tradition. *Arctic*. 54(4): 377–393.

Cullenward, D. and Victor, D.G. (2020) *Making Climate Policy Work*. Cambridge: Polity Press. 256pp.

Culver, L. (2014) Seeing climate through culture. *Environmental History*. 19(2): 311–318.

D'Alisa, G., Demaria, F. and Kallis, G. (eds.) (2014) *Degrowth: A Vocabulary for a New Era*. Abingdon: Routledge.

Dalton, A.M. (2018) Eco-theology. Chapter 3.2 (pp. 271–274) In: *Companion to Environmental Studies*. Castree, N., Hulme, M. and Proctor, J.D. (eds.). Abingdon: Routledge.

Daoudy, M. (2020) *The Origins of the Syrian Conflict: Climate Change and Human Security*. Cambridge: Cambridge University Press.

Dasgupta, P. and Ramanathan, V. (2014) Pursuit of the common good. *Science*. 345: 1457–1458.

Daston, L. (1991) Marvelous facts and miraculous evidence in early modern Europe. *Critical Inquiry*. 18: 93–124.

Daston, L. (2010) The world in order. (pp. 15–34) In: *Without Nature? A New Condition for Theology*. Alberston, D. and King, C. (eds.). Bronx, NY: Fordham University Press.

Davies, W. (2019) *Green Populism? – Action and Mortality in the Anthropocene*. University of Surrey: Centre for Understanding Sustainable Prosperity.

Deane-Drummond, C. (2008) *Eco-theology*. London: Darton, Longman and Todd.

Death, C. (2011) Summit theatre: Exemplary governmentality and environmental diplomacy in Johannesberg and Copenhagen. *Environmental Politics*. 20(1): 1–19.

Degroot, D. (2018) *The Frigid Golden Age: Climate Change, the Little Ice Age, and the Dutch Republic, 1560–1720*. Cambridge: Cambridge University Press.

Delingpole, J. (2012) *Watermelons: How Environmentalists are Killing the Planet, Destroying the Economy and Stealing your Children's Future*. London, UK: Biteback Books.

Diamond, J. (1997) *Guns, Germs and Steel: The Fates of Human Societies*. New York: W.W. Norton.

Dingwerth, K. and Green, J.F. (2015) Transnationalism. Chapter 14 (pp. 3153–163) In: *Research Handbook on Climate Governance*. Bäckstrand, K. and Lövbrand, E. (eds.). Cheltenham: Edward Elgar Publishing.

Dolgoy, E.A. (2016) The scientific and the scientistic: Roger Scruton on the consequences of modern science. *Perspectives on Political Science*. 45(4): 244–250.

Dotson, T. (2021) *The Divide: How Fanatical Certitude Is Destroying Democracy*. Cambridge, MA: MIT Press.

Douglas, R. (2018) *The Commonplaces of Environmental Scepticism*. Centre for the Understanding of Sustainable Prosperity (CUSP) Working Paper No. 17. Guildford: University of Surrey.

Doyle, J. (2007) Picturing the clima(c)tic: Greenpeace and the representational politics of climate change communication. *Science as Culture*. 16(2), 129–150.

Dry, S. (2019) *Waters of the World: The Story of the Scientists who Unraveled the Mysteries of our Seas, Glaciers and Atmosphere – and Made the Planet Whole*. Chicago, IL: University of Chicago Press.

Dunlap, R.E. and Jacques, P.J. (2013) Climate change denial books and conservative think tanks: Exploring the connection. *American Behavioural Scientist*. 57(6): 699–731.

Dwyer, C. (2016) Why does religion matter for cultural geographers? *Social & Cultural Geography*. 17(6): 758–762.

Eckersley, R. (2004) *The Green State: Rethinking Democracy and Sovereignty*. Cambridge, MA: MIT Press.

Edwards, P.N. (2006) Meteorology as infrastructural globalism. *Osiris*. 21: 229–250.

Edwards, P.N. (2010) *A Vast Machine: Computer Models, Climate Data and the Politics of Global Warming*. Cambridge, MA: MIT Press.

Elvin, M. (1998) Who was responsible for the weather? Moral meteorology in late imperial China. *Osiris*. 13: 213–237.

Endfield, G.H. (2019) Weather and elemental places. *Historical Geography*. 47: 1–31.

Erickson, P., Klein, J.L., Daston, L., Lemov, R., Sturm, T. and Gordin, M.D. (2013) *How Reason Almost Lost Its Mind. The Strange Career of Cold War Rationality*. Chicago, IL: University of Chicago Press.

Evans, A. (2017) *The Myth Gap: What Happens When Evidence and Arguments Aren't Enough?* London: Eden Books.

Fair, H. (2018) Three stories of Noah: Navigating religious climate change narratives in the Pacific Island region. *Geo: Geography and Environment*. 5(2): e00068.

Farbotko, C. (2005) Tuvalu and climate change: Constructions of environmental displacement in the Sydney Morning Herald. *Geografiska Annaler*. 87B(4): 279–293.

Fischer, F. (2017) *Climate Crisis and the Democratic Prospect: Participatory Governance in Sustainable Communities*. Oxford: Oxford University Press.

Fisher, D.R. (2019) The broader importance of #FridaysForFuture. *Nature Climate Change*. 9(6): 430–431.

Fleming, J.R. (2010) *Fixing the Sky: The Checkered History of Weather and Climate Control*. New York: Columbia University Press.

Fleming, J.R. and Jankovic, V. (eds.) (2011) Klima. *Osiris*. 26: 350.

Flohn, H. (1977) Climate and energy: A scenario to a 21st century problem. *Climatic Change*. 1: 5–20.

Forchtner, B. (2019) Climate change and the far right. *WIREs Climate Change*. 10(5): e604.

Ford, T.H. (2016) Climate change and literary history. Chapter 9 (pp. 157–174) In: *A Cultural History of Climate Change*. Bristow, T. and Ford, C.H. (eds.). Abingdon: Routledge.

Foster, J.B. (1999) Marx's theory of metabolic rift: Classical foundation for environmental sociology. *American Journal of Sociology*. 105(2): 366–405.

Frisch, M. (1980) *Man in the Holocene*. London: Eyre, Methuen.

Gagne, K. (2019) *Caring for Glaciers: Land, Animals, and Humanity in the Himalayas*. Seattle, WA: University of Washington Press.

Galafassi, D., Kagan, S., Milkoreit, M. and co-authors (2018) 'Raising the temperature': The arts on a warming planet. *Current Opinion in Environmental Sustainability*. 31: 71–79.

Gallo, I. (2018) Between Kasache and Geneva: The multi-sited voyage of climate-resilient development in Malawi. Unpublished PhD Thesis. Lancaster, UK: University of Lancaster.

Gardner, C.J. and Wordley, C.F.R. (2019) Scientists must act on our own warnings to humanity. *Nature Ecology & Evolution*. 3(9): 1271–1272.

Garrard, G., Goodbody, A., Handley, G. and Posthumus, S. (2019) *Climate Change Scepticism: A Transnational Ecocritical Analysis*. London: Bloomsbury Academic.

Gay-Antaki, M. and Liverman, D. (2018) Climate for women in climate science: Women scientists and the IPCC. *Proceedings of the National Academy of Sciences*. 115(9): 2060–2065.

Geden, O. (2015) Climate advisors must maintain integrity. *Nature*. 521: 27–28.

Genus, A. and Stirling, A. (2018) Collingridge and the dilemma of control: Towards responsible and accountable innovation. *Research Policy*. 47(1): 61–69.

Geoghegan, H. and Leyson, C. (2012) On climate change and cultural geography: Farming on the Lizard Peninsula, Cornwall, UK. *Climatic Change*. 113(1): 55–66.

Germani, F. (2019) Greta Thunberg is turning revolutionary and why this isn't any good. *Culturico*. 26 October. Available at https://culturico.com/2019/10/26/greta-thunberg-is-turning-revolutionary-and-why-this-isnt-any-good/ [Accessed 22 April 2020].

Ghosh, A. (2016) *The Great Derangement: Climate Change and the Unthinkable*. Chicago, IL: University of Chicago Press.

Ghosh, U., Bose, S. and Bramhachari, R. (2018) *Living on the Edge: Climate Change and Uncertainty in the Indian Sundarbans*, STEPS Working Paper 101. Brighton: STEPS Centre.

Giddens, A. (1998) *Beyond Left and Right: The Future of Radical Politics*. Cambridge: Polity Press.

Giddens, A. (2009) *The Politics of Climate Change*. London: Polity Press.

Gieryn, T.F. (2006) City as truth-spot: Laboratories and field-sites in urban studies. *Social Studies of Science*. 36(1): 5–38.

Gilman, N. (2020) The coming avocado politics: What happens when the ethno-nationalist right gets serious about the climate emergency. *The Breakthrough Journal*. 12(Winter).

Glacken, C. (1967) *Traces on a Rhodian Shore: Nature and Culture in Western Thought from Ancient Times to the End of the Eighteenth Century*. Berkeley, CA: University of California Press. 761pp.

Goodbody, A. and Johns-Putra, A. (eds.) (2018) *Cli-Fi: A Companion*. Oxford: Peter Lang.

Goodbody, A. and Johns-Putra, A. (2019) The rise of the climate change novel. Chapter 14 (pp. 229–246) In: *Climate and Literature*. Johns-Putra, A. (ed.). Cambridge: Cambridge University Press.

Griffiths, M. (2017) *The New Poetics of Climate Change: Modernist Aesthetics for a Warming World*. London: Bloomsbury Academic.

Griffiths, T. (2007) The humanities and an environmentally sustainable Australia. *Australian Humanities Review*. 43.

Grim, J. and Tucker, M.E. (2014) *Ecology and Religion: Foundations of Contemporary Environmental Studies Series*. Washington, DC: Island Press.

Grosz, E. (1999) Becoming . . . an introduction. (pp. 1–11) In: *Becomings: Explorations in Time, Memory and Future*. Grosz, E. (ed.). Ithaca, NY: Cornell University Press.

Grove, R.H. (1998) *Ecology, Climate and Empire: The Indian Legacy in Global Environmental History, 1400–1940*. Delhi, India: Oxford University Press.

Grove, R.H. and Adamson, G.C.D. (2018) *Niño and World History*. London: Palgrave Macmillan.

Gupta, J. (2014) *The History of Global Climate Governance*. Cambridge: Cambridge University Press.

Guston, D.H., Finn, E. and Robert, J.S. (eds.) (2017) *Frankenstein. Annotated for Scientists, Engineers and Creators of All Kinds*. Cambridge, MA: MIT Press.

Habermas, J. (1984) *The Theory of Communicative Action*, vol. 1. Cambridge: Polity Press.

Habermas, J., Reder, M., Brieskorn, N., Ricken, F., and Schmidt, J. (2010) *An Awareness of What Is Missing: Faith and Reason in a Post-secular Age*. Cambridge: Polity Press.

Haites, E. (2018) Carbon taxes and greenhouse gas emissions trading systems: What have we learned? *Climate Policy*. 18(8): 955–966.

Hannah, D. (ed.) (2018) *A Year Without a Winter*. New York: Columbia University Press.

Hansman, H. (2015) This song is composed from 133 years of climate change data. *Smithsonian Magazine*. 21 September. Available at www.smithsonianmag.com/.

Haraway, D. (1988) Situated knowledges: The science question in feminism and the privilege of partial perspective. *Feminist Studies*. 14(3): 575–599.

Hardwick, J. and Stephens, R.J. (2020) Acts of god: Continuities and change in Christian responses to extreme weather events from early modernity to the present. *WIREs Climate Change*. 11(2): e631.

Hart, D.M. and Victor, D.G. (1993) Scientific elites and the making of US policy for climate change research, 1957–1974. *Social Studies of Science*. 23(4): 643–680.

Hathaway, J.R. (2020) Climate change, the intersectional imperative, and the opportunity of the green new deal. *Environmental Communication*. 14(1): 13–22.

Hausfather, Z. and Peters, G.P. (2020) Emissions – The 'business as usual' story is misleading. *Nature*. 577: 618–620.

Hawkins, H. and Kanngieser, A. (2017) Artful climate change communication: Overcoming abstractions, insensibilities, and distances. *WIREs Climate Change*. 8(5): e472.

Hayward, B., Salili, D.H., Tupuana'i, L.L. and Tualamali'i', J. (2020) It's not "too late": Learning from Pacific small island developing states in a warming world. *WIREs Climate Change*. 11(1): e612.

Heaney, S. (2006) *District and Circle*. London: Faber & Faber.

Heidenreich, F. (2018) How will sustainability transform democracy? Reflections on an important dimension of transformation sciences. *GAIA*. 27(4): 357–362.

Heikkinen, M., Ylä-Anttila, T. and Juhola, S. (2019) Incremental, reformistic or transformational: What kind of change do C40 cities advocate to deal with climate change? *Journal of Environmental Policy & Planning*. 21(1): 90–103.

Heise, U.K. (2008) *Sense of Place and Sense of Planet*. Oxford: Oxford University Press.

Heise, U.K. (2016) *Imagining Extinction: The Cultural Meanings of Endangered Species*. Chicago, IL: Chicago University Press.

Henson, R. (2006) *The Rough Guide to Climate Change*. London: Rough Guides Ltd.

Herbertson, A.J. (1905) The major natural regions: An essay in systematic geography. *The Geographical Journal*. 25(3): 300–310.

Heymann, M. and Dahan-Dalmedico, A. (2019) Epistemology and politics in earth system modeling: Historical perspectives. *Journal of Advances in Modeling Earth Systems*. 11(5): 1139–1152.

Hibbard, K., Wilson, T., Averyt, K., Harriss, R., Newmark, R., Rose, S., Shevliakova, E. and Tidwell, V. (2014) Energy, water, and land use. Chapter 10 (pp. 257–281) In: *Climate Change Impacts in the United States: The Third National Climate Assessment*. Melillo, J.M., Richmond, T.C. and Yohe, G.W. (eds.). Washington, DC: U.S. Global Change Research Program.

Higgins, D. (2017) *British Romanticism, Climate Change and the Anthropocene: Writing Tambora*. London: Palgrave Macmillan.

Higgins, D. (2019) British romanticism and the global climate. Chapter 8 (pp. 128–143) In: *Climate and Literature*. Johns-Putra, A. (ed.). Cambridge: Cambridge University Press.

Hildebrandsson, H.H. and de Bort, T. (1896) *Atlas International des Nuages: Pub Conformenent aux Decisions du Comite Meteorologique International*. Paris: Gauthier-Villars.

Hobson, K. and Niemeyer, S. (2013) "What sceptics believe": The effects of information and deliberation on climate change skepticism. *Public Understanding of Science*. 22(4): 396–412.

Holgate, B. (2019) *Climate and Crises: Magical Realism as Environmental Discourse*. London and New York: Routledge.

Hopkins, R. (2008) *The Transition Handbook: From Oil Dependency to Local Resilience*. Stroud: Green Books.

Horton, J. and Keith, D. (2016) Solar geoengineering and obligations to the global poor. (pp. 79–92) In: *Climate Justice and Geoengineering: Ethics and Policy in the Atmospheric Anthropocene*. Preston, C.J. (ed.). Lanham, MD: Rowman & Littlefield International.

Houser, H. (2020) *Infowhelm: Environmental Art and Literature in an Age of Data*. New York: Columbia University Press.

Howarth, C.C. and Sharman, A.G. (2015) Labeling opinions in the climate debate: A critical review. *WIREs Climate Change*. 6(2): 239–254.

Howlett, P. and Morgan, M.S. (2010) *How Well Do Facts Travel? The Dissemination of Reliable Knowledge*. Cambridge: Cambridge University Press.

Hsiang, S.M., Burke, M. and Miguel, E. (2013) Quantifying the influence of climate on human conflict. *Science*. 341: 1212.

Huang, Y. (2017) Confucianism. Chapter 6 (pp. 52–59) In: *Routledge Handbook on Religion and Ecology*. Jenkins, W., Tucker, M.E. and Grim, J. (eds.). Abingdon: Routledge.

Hughes, H.R. and Paterson, M. (2017) Narrowing the climate field: The symbolic power of authors in the IPCC's assessment of mitigation. *Review of Policy Research*. 34(6): 744–766.

Hulme, K. (1984) *The Bone People*. Christchurch, New Zealand: The Spiral Press.

Hulme, M. (2009) *Why We Disagree About Climate Change: Understanding Controversy, Inaction and Opportunity*. Cambridge: Cambridge University Press.

Hulme, M. (2010) Problems with making and governing global kinds of knowledge. *Global Environmental Change*. 20(4): 558–564.

Hulme, M. (2011) Reducing the future to climate: A story of climate determinism and reductionism. *Osiris*. 26(1): 245–266.

Hulme, M. (2013) How climate models gain and exercise authority. Chapter 2 (pp. 30–44) In: *The Social Life of Climate Change Models: Anticipating Nature*. Hastrup, K. and Skrydstrup, M. (eds.). Abingdon: Routledge.

Hulme, M. (2014a) Climate change and virtue: An apologetic. *Humanities*. 3(3): 299–312.

Hulme, M. (2014b) *Can Science Fix Climate Change? A Case Against Climate Engineering*. Cambridge, UK: Polity Press.

Hulme, M. (2016) *Weathered: Cultures of Climate*. London: Sage.

Hulme, M. (2017) Climate change and the significance of religion. *Economic & Political Weekly*. 52(28): 14–17.

Hulme, M. (2018) Weather-worlds of the Anthropocene and the end of climate. *Weber: The Contemporary West*. Fall Issue: 63–74.

Hulme, M. (ed.) (2020) *Contemporary Climate Change Debates: A Student Primer*. Abingdon: Routledge.

Hunt, J.D., Nascimento, A., Diuana, F.A. and co-authors (2020) Cooling down the world oceans and the earth by enhancing the North Atlantic Ocean current. *SN Applied Sciences*. 2: 15.

Huntington, E.W. (1915) *Civilization and Climate* (1st ed.). New Haven, CT: Yale University Press.

Ingold, T. (1994) Introduction to culture. Chapter 12 (pp. 329–349) In: *Companion Encyclopedia of Anthropology: Humanity, Culture and Social Life*. Ingold, T. (ed.). London: Routledge.

Interfaith Summit on Climate Change (2014) Climate, faith and hope: Faith traditions together for a common future. Available at https://unfccc.int/files/meetings/lima_dec_2014/statements/application/pdf/cop20_hls_faith.pdf [Accessed 2 September 2020].

IPCC (2018) *Global Warming of 1.5°C: An IPCC Special Report*. Geneva: IPCC.

Jacka, J. (2009) Correlating local knowledge with climatic data: Porgeran experiences of climate change in Papua New Guinea. Chapter 9 In: *Anthropology and Climate: From Encounters to Actions*. Crate, S.A. and Nuttall, M. (eds.). Walnut Creek, CA: Left Coast Press.

Jackson, T. (2017) *Prosperity Without Growth: Foundations for the Economy of Tomorrow* (2nd ed.) Abingdon, UK: Routledge.

Jacobson, M.Z. (2020) *100% Clean, Renewable Energy and Storage for Everything*. Cambridge: Cambridge University Press.

Jamison, A. (2010) Climate change knowledge and social movement theory. *WIREs Climate Change*. 1(6): 811–823.

Jankovic, V. (2000) *Reading the Skies: A Cultural History of English Weather, 1650–1820*. Manchester: Manchester University Press.

Jasanoff, S. (2005) *Designs on Nature: Science and Democracy in Europe and the United States*. Princeton, NJ: Princeton University Press.

Jasanoff, S. (2010) A new climate for society. *Theory, Culture & Society*. 27(2/3): 233–253.

Jenkins, W., Berry, E. and Kreider, L.B. (2018) Religion and climate change. *Annual Review of Environment and Resources*. 43: 9.1–9.24.

Johnson, D. (2019) *Watsuji on Nature: Japanese Philosophy in the Wake of Heidegger*. Evanston, IL: Northwestern University Press.

Jones, P.D., New, M., Parker, D.E., Martin, S. and Rigor, I.G. (1999) Surface air temperature and its changes over the past 150 years. *Reviews of Geophysics*. 37(2): 173–199.

Jones, R.P., Cox, D. and Navarro-Rivera, J. (2014) *Believers, Sympathizers and Skeptics: Why Americans are Conflicted about Climate Change, Environmental Policy and Science*. Washington, DC: Public Religion Research Institute (PRRI).

Kagan, J. (2009) *The Three Cultures: Natural Sciences, Social Sciences and the Humanities in the 21st Century*. Cambridge: Cambridge University Press.

Kahan, D., Jenkins-Smith, H. and Braman, D. (2011) Cultural cognition of scientific consensus. *Journal of Risk Research*. 14(2): 147–174.

Kallis, G. (2011) In defense of degrowth. *Ecological Economics*. 70: 873–880.

Karlsson, R. (2016) Three metaphors for sustainability in the Anthropocene. *Anthropocene Review*. 3(1): 23–32.

Keighren, I.M. (2010) *Bringing Geography to Book: Ellen Semple and the Reception of Geographical Knowledge*. London: I B Tauris.

Kelly, D. (2019) *Politics and the Anthropocene*. Cambridge: Polity Press.

Kempf, W. (2015) Representation as disaster: Mapping islands, climate change and displacement in Oceania. *Pacific Studies*. 38(1–2): 200–228.

Keohane, R.O. and Victor, D.G. (2011) The regime complex for climate change. *Perspectives on Politics*. 9(1): 7–23.

Kerber, G. (2010) Caring for creation and striving for climate justice: Implications for mission and spirituality. *International Review of Mission*. 99(2): 219–229.

Kidwell, J., Ginn, F., Northcott, M., Bomberg, E. and Hague, A. (2018) Christian climate care: Slow change, modesty and eco-theo-citizenship. *Geo: Geography and Environment*. 5(2): e00059.

Kincer, J.B. (1933) Is our climate changing? A study of long-term temperature trends. *Monthly Weather Review*. 61: 251–259.

King, D., Schrag, D., Dadi, Z., Ye, Q. and Ghosh, A. (2015) *Climate Change: A Risk Assessment*. London: Foreign & Commonwealth Office.

Kingsolver, B. (2012) *Flight Behaviour: A Novel*. London: Faber & Faber.

Klein, N. (2014) *This Changes Everything: Capitalism vs the Climate*. New York: Simon & Schuster.

Klintman, M. and Boström, M. (2015) Citizen-consumers. Chapter 47 (pp. 309–319) In: *Research Handbook on Climate Governance*. Bäckstrand, K. and Lövbrand, E. (eds.). Cheltenham: Edward Elgar Publishing.

Kneale, J. and Randalls, S. (2014) Invisible atmospheric knowledges in British insurance companies, 1830–1914. *History of Meteorology*. 6: 35–52.

Knutti, R. (2010) The end of model democracy? *Climatic Change*. 102(3–4): 395–404.

Knutti, R. and Sedláček, J. (2013) Robustness and uncertainties in the new CMIP5 climate model projections. *Nature Climate Change*. 3(4): 369–373.

Koehrsen, J. (2021) Islam and climate change. *WIREs Climate Change*. 12(3): e702.

Koster, H.P. and Conradie, E.M. (eds.) (2019) *T&T Clark Companion of Christian Theology and Climate Change*. London: Bloomsbury.

Kupperman, K.O. (1982) The puzzle of the American climate in the early colonial period. *The American Historical Review*. 87(5): 1262–1289.

Lahsen, M. (2013) Anatomy of dissent: A cultural analysis of climate scepticism. *American Behavioural Scientist*. 57(6): 732–753.

Lamb, H.H. (1971) Climate-Engineering schemes to meet a climatic emergency. *Earth-Science Reviews*. 7: 87–95.

Lamb, H.H. (1972) *Climate: Past, Present and Future. Vol. I: Fundamentals and Climate Now*. London: Methuen.

Lamb, H.H. (1977) *Climate: Past, Present and Future. Vol. II: Climatic History and the Future*. London: Methuen.

Lamb, H.H. (1982) *Climate, History and the Modern World*. London: Methuen.

Larson, B. (2011) *Metaphors for Environmental Sustainability: Redefining Our Relationship with Nature*. New Haven, CT: Yale University Press.

Latour, B. (2011) Love your monsters: Why we must care for our technologies as we do our children. *Breakthrough Journal*. 2: Fall Issue.

Latour, B. (2018) *Down to Earth: Politics in the New Climatic Regime*. Cambridge, UK: Polity Press.

Lawrence, M.G. and Schäfer, S. (2019) Promises and perils of the Paris agreement. *Science*. 364: 829–830.

Leduc, T.B. (2010) *Climate, Culture, Change: Inuit and Western Dialogues with a Warming North*. Ottawa, Ontario: University of Ottawa Press.

Lejano, R. (2019) Ideology and the narrative of climate scepticism. *Bulletin of the American Meteorological Society*. 100(12): ES415–ES421.

Lidström, S. and Garrard, G. (2014) "Images adequate to our predicament": Ecology, environment and ecopoetics. *Environmental Humanities*. 5: 35–53.

Livingstone, D.N. (2003) *Putting Science in its Place: Geographies of Scientific Knowledge*. Chicago, IL: University of Chicago Press.

Livingstone, D.N. (2005) Science, text and space: Thoughts on the geography of reading. *Transactions of the Institute of British Geographers*. 30(4): 391–401.

Locher, F. and Fressoz, J.B. (2012) The frail climate of modernity. A climate history of environmental reflexivity. *Critical Inquiry*. 38(3): 579–598.

Lomborg, B. (2001) *The Skeptical Environmentalist: Measuring the Real State of the World*. Cambridge: Cambridge University Press.

Lomborg, B. (ed.) (2004) *Global Crises, Global Solutions*. Cambridge: Cambridge University Press.

Long, J. (2020) Is it necessary to research solar climate engineering as a possible backstop technology? YES: Research gives society an opportunity to act responsibly. Chapter 8 (pp. 109–114) In: *Contemporary Climate Change Debates: A Student Primer*. Hulme, M. (ed.). Abingdon: Routledge.

Lovelock, J. (1979) *Gaia, A New Look at Life on Earth*. Oxford: Oxford University Press.

Lucas, C.H. and Davison, A. (2019) Not 'getting on the bandwagon': When climate change is a matter of unconcern. *Environment & Planning E: Nature & Space*. 2(1): 129–149.

Luke, T.W. (2018) Tracing race, ethnicity and civilization in the Anthropocene. *Society and Space*. 38(1): 129–146.

MacIntyre, A. (1981) *After Virtue: A Study in Moral Theory*. Notre Dame, IN: University of Notre Dame Press.

Maggs, D. and Robinson, J. (2020) *Sustainability in an Imaginary World: Art and the Question of Agency*. Abingdon: Routledge.

Mahony, M. (2014) The predictive state: Science, territory and the future of the Indian climate. *Social Studies of Science*. 44(1): 109–133.

Mahony, M. (2016a) Empire for an empire of 'all types of climate': Meteorology as an imperial. *Journal of Historical Geography*. 51: 29–39.

Mahony, M. (2016b) Picturing the future-conditional: Montage and the global geographies of climate change. *Geo: Geography and Environment*. 3(2): e00019.

Mahony, M. and Endfield, G. (2018) Climate and colonialism. *WIREs Climate Change*. 9(2): e510.

Mahony, M. and Hulme, M. (2018) Epistemic geographies of climate change: Science, space and politics. *Progress in Human Geography*. 42(3): 395–424.

Mahony, M. and Randalls, S. (eds.) (2020) *Weather, Climate and the Geographical Imagination: Placing Atmospheric Knowledges*. Pittsburgh, PA: University of Pittsburgh Press.

Mair, S. and Steinberger, J. (2019) In the age of extinction, who is extreme? – A response to policy exchange's "extremism rebellion" report. 23 July. Available at www.opendemocracy.net/en/oureconomy/age-extinction-who-extreme-response-policy-exchange/

Malm, A. (2016) *Fossil Capital: The Rise of Steam Power and the Roots of Global Warming*. London: Verso Books.

Mann, M.E. (2012) *The Hockey Stick and the Climate Wars: Dispatches from the Front Lines*. New York: Columbia University Press.

Markley, R. (2019) Literature, climate and time: Between history and story. Chapter 1 (pp. 15–23) In: *Climate and Literature*. Johns-Putra, A. (ed.). Cambridge: Cambridge University Press.

Marris, E. (2011) *Rumbunctious Garden: Saving Nature in a Post-Wild World*. London: Bloomsbury. 210pp.

Marshall, T. (2015) *Prisoners of Geography: Ten Maps That Tell You Everything You Need to Know About Global Politics*. London: Elliot & Thompson.

Masco, J. (2010) Bad weather: Or planetary crisis. *Social Studies of Science*. 40(1): 7–40.
Mazower, M. (2012) *Governing the World: The History of an Idea*. London: Penguin Books.
Mazzucatu, M. (2013) *The Entrepreneurial State: Debunking Public vs. Private Sector Myths*. London: Anthem Press.
Mazzucatu, M. (2021) *Mission Economy: A Moonshot Guide to Changing Capitalism*. London: Allen Lane.
McCarthy, C. (2006) *The Road*. London: Picador.
McEwan, I. (2010) *Solar*. London: Jonathan Cape.
McLaren, D.P. (2018) Whose climate and whose ethics? Conceptions of justice in solar geoengineering modelling. *Energy Research & Social Science*. 44: 209–221.
McLaren, D.P. and Markusson, N. (2020) The co-evolution of technological promises, modelling, policies and climate change targets. *Nature Climate Change*. 10(5): 392–397.
Meadows, D.H., Meadows, D.L., Randers, J. and Behrens III, W.W. (1972) *The Limits to Growth; A Report for the Club of Rome's Project on the Predicament of Mankind*. New York: Universe Books.
Meckling, J. and Allan, B.B. (2020) The evolution of ideas in global climate policy. *Nature Climate Change*. 10(5): 434–438.
Melanesian Mission (2019) The Anglican church of Melanesia and climate change. 25 November. Available at www.mmuk.net/news/the-anglican-church-of-melanesia-and-climate-change/ [Accessed 14 August 2020].
Mercer, H. (2020) How newcomers tried to understand the climate of colonial Australia, c. 1760s – 1860s. Unpublished PhD Thesis. University of Oxford, Oxford.
Miglietti, S. (2022) *The Empire of Climate: Early Modern Climate Theories and the Problem of Human Agency*. Cambridge: Cambridge University Press.
Miglietti, S. and Morgan, J. (eds.) (2017) *Governing the Environment in the Early Modern World: Theory and Practice*. Abingdon: Routledge.
Milkoreit, M. (2016) The promise of climate fiction: Imagination, storytelling and the politics of the future. Chapter 10 (pp. 171–191) In: *Reimagining Climate Change*. Wapner, P. and Elver, H. (eds.). Abingdon: Routledge.
Ming-Yi, W. (2013) *The Man with Compound Eyes*. Taiwan: Summer Festival Press.
Montford, A.W. (2012) *Nullius in Verbia: On the Word of No-One*. London: Global Warming Policy Foundation.
Monthly Weather Review (1911) 106 climatological sections in the united states. *Monthly Weather Review*. 39(2): Supplemental Charts. https://doi.org/10.1175/1520-0493-39.2.s.
More, T. (1516/1967) *Utopia*. (trans. Dolan, J.P.). New York: New American Library.
Morgan, R.A. (2020) Looking for the Leeuwin: An environmental history of the Leeuwin Current. Chapter 5 (pp. 93–113) In: *Weather, Climate and the Geographical Imagination: Placing Atmospheric Knowledges*. Mahony, M. and Randalls, S. (eds.). Pittsburgh, PA: University of Pittsburgh Press.
Morton, T. (2013) *Hyperobjects: Philosophy and Ecology After the End of the World*. Minneapolis, MN: University of Minnesota Press.
Mouffe, C. (2005) *On the Political*. Abingdon: Routledge.
Nanda, M. (2003) *Prophets Facing Backwards: Post-modern Critiques of Science and Hindu Nationalism in India*. New Brunswick, NJ: Rutgers University Press.

Nemet, G. (2020) Should future investments in energy technology be limited exclusively to renewables? NO: A diverse clean energy portfolio delivers wide social and economic benefits. Chapter 7 (pp. 101–107) In: *Contemporary Climate Change Debates: A Student Primer*. Hulme, M. (ed.). Abingdon: Routledge.

Nemeth, C. (2018) *In Defense of Troublemakers: The Power of Dissent in Life and Business*. New York: Basic Books.

Nerlich, B. and Jaspal, R. (2012) Metaphors we die by? Geoengineering, metaphors and the argument from catastrophe. *Metaphor and Symbol*. 27(2): 131–147.

Neumayer, E. (2007) A missed opportunity: The stern review on climate change fails to tackle the issue of non-substitutable loss of natural capital. *Global Environmental Change*. 17(3–4): 297–301.

Newell, J., Robin, L. and Wehner, K. (eds.) (2016) *Curating the Future: Museums, Communities and Climate Change*. Abingdon: Routledge.

Nightingale, A.J., Eriksen, S., Taylor, M., Forsyth, T., Pelling, M. and co-authors (2020) Beyond technical fixes: Climate solutions and the great derangement. *Climate and Development*. 12(4): 343–352.

Nordhaus, T. (2019) Eco-identity politics: An editorial. *Breakthrough Journal*. 10(February).

Nordhaus, T. and Shellenberger, M. (2007) *Break Through: From the Death of Environmentalism to the Politics of Possibility*. Boston, MA: Houghton Miflin.

North, P. (2011) The politics of climate activism in the UK: A social movement analysis. *Environment and Planning A*. 43: 1581–1598.

Northcott, M.S. (2007) *Moral Climate: The Ethics of Global Warming*. London: Dartman, Longman and Todd.

Norton, C. and Hulme, M. (2019) Telling one story, or many? An ecolinguistic analysis of climate change stories in UK national newspaper editorials. *Geoforum*. 104: 114–136.

NRC (1986) *Earth System Science. Overview: A Program for Global Change*. Washington, DC: National Academies Press.

Nurmis, J. (2016) Visual climate change art 2005–2015: Discourse and practice. *WIREs Climate Change*. 6(4): 501–516.

Obermeister, N. and Honybun-Arnolda, E. (2019) Civilisation VI: Gathering Storm shows video games can make us think seriously about climate change. *The Conversation*. 15 February. Available at https://theconversation.com/civilization-vi-gathering-storm-shows-video-games-can-make-us-think-seriously-about-climate-change-111791.

O'Brien, C. (2016) Rethinking seasons: Changing climate, changing time. Chapter 2 (pp. 38–54) In: *A Cultural History of Climate Change*. Bristow, T. and Ford, C.H. (eds.). Abingdon: Routledge.

Oldfield, J.D. (2013) Climate modification and climate change debates among Soviet physical geographers, 1940s – 1960s. *WIREs Climate Change*. 4(6): 513–524.

O'Neill, B.C., Kriegler, E., Ebi, K.L., Kemp-Benedict, E., Riahi, K. and Rothman, D.S. (2017) The roads ahead: Narratives for shared socioeconomic pathways describing world futures in the 21st century. *Global Environmental Change*. 42: 169–180.

O'Neill, D.W. (2012) Measuring progress towards a socially sustainable steady state economy. Unpublished PhD Thesis. Leeds, UK: University of Leeds.

O'Neill, K. (2009) *The Environment and International Relations*. Cambridge: Cambridge University Press.

Operation Noah (2012) Climate change and the purposes of god: A call to the church. Available at http://operationnoah.org/what-we-do/ash-wednesday-declaration/ [Accessed 13 August 2020].

Oreskes, N. and Conway, E.M. (2010) *Merchants of Doubt: How a Handful of Scientists Obscured the Truth on Issues from Tobacco Smoke to Global Warming*. London: Bloomsbury.

Oreskes, N. and Conway, E.M. (2014) *The Collapse of Western Civilisation: A View from the Future*. New York: Columbia University Press.

Orwell, G. (1949) *Nineteen Eighty Four: A Novel*. London: Secker & Warburg.

Ostrom, E. (1990) *Governing the Commons: The Evolution of Institutions for Collective Action*. Cambridge: Cambridge University Press.

Paerregaard, K. (2020) Communicating the inevitable: Climate awareness, climate discord and climate research in Peru's highland communities. *Environmental Communication*. 14(1): 112–125.

Parker, G. (2013) *Global Crisis: War, Climate Change and Catastrophe in the Seventeenth Century*. New Haven, CT: Yale University Press.

Pasisi, J. (2019) Cultural resilience and climate change: Everyday lives in Niue. (pp. 168–174) In: *Climate Futures Re-Imagining Global Climate Justice*. Bhavnani, K.K., Foran, J., Kurian, P.A. and Munshi, D. (eds.). London: Zed Books.

Pearse, R. and Böhm, S. (2014) Ten reasons why carbon markets will not bring about radical emissions reduction. *Carbon Management*. 5(4): 325–337.

Pettifor, A. (2019) *The Case for the Green New Deal*. London and Brooklyn: Verso Books.

Pezzey, J.C.V. (2019) Why the social cost of carbon will always be disputed. *WIREs Climate Change*. 10(1): e558.

Pinker, S. (2019) *Enlightenment Now: The Case for Reason, Science, Humanism and Progress*. London: Penguin Random House.

Pirgmaier, E. (2018) Marx for environmentalists: Rise up! speak up! insist! *GAIA*. 27(3): 265.

Poortinga, W., Spence, A., Whitmarsh, L., Capstick, S. and Pidgeon, N.F. (2011) Uncertain climate: An investigation into public scepticism about anthropogenic climate change. *Global Environmental Change*. 21(3): 1015–1024.

Pope Francis (2015) *Encyclical letter Laudato Si' of the Holy Father Francis – On Care for Our Common Home*. Rome: Vatican Press.

Qin, T. and Zhang, M. (2020) Does the 'Chinese model' of environmental governance demonstrate to the world how to govern the climate? YES: It offers a centralised governance model with a command and control approach. Chapter 14 (pp. 207–212) In: *Contemporary Climate Change Debates: A Student Primer*. Hulme, M. (ed.). Abingdon: Routledge.

Rabinow, P. (ed.) (1991) *The Foucault Reader: An Introduction to Foucault's Thought*. London: Penguin.

Ramírez-i-Ollé, M. (2015) Rhetorical strategies for scientific authority: A boundary-work analysis of 'Climategate'. *Science as Culture*. 24(4): 384–411.

Ramírez-i-Ollé, M. (2018) 'Civil skepticism' and the social construction of knowledge: A case in dendroclimatology. *Social Studies of Science*. 48(6): 821–845.

Rawls, J. (2005) *Political Liberalism*. New York: Columbia University Press.

Raworth, K. (2018) *Doughnut Economics: Seven Ways to Think Like a 21st-Century Economist*. London: Random House.

Reisz, M. (2019) *Too Little, Too Late? A Profile of Jem Bendell*. London: Times Higher Education. 12 September, p. 35.

Renn, J. (2020) *The Evolution of Knowledge: Rethinking Science for the Anthropocene*. Princeton, NJ: Princeton University Press.

Revelle, R. and Suess, H.E. (1957) Carbon dioxide exchange between atmosphere and ocean and the question of an increase of atmospheric CO2 during the past decades. *Tellus*. 9: 18–27.

Reynolds, J.L. (2019) *The Governance of Solar Geoengineering: Managing Climate Change in the Anthropocene*. Cambridge: Cambridge University Press.

Rice, J.L., Burke, B.J. and Heynen, N. (2015) Knowing climate change, embodying climate praxis: Experiential knowledge in Southern Appalachia. *Annals of the Association of American Geographers*. 105(2): 253–262.

Rickards, L.A. (2015) Metaphor and the Anthropocene: Presenting humans as a geological force. *Geographical Research*. 53(3): 280–287.

Ridley, M. (2015) *The Climate Wars and the Damage to Science*. London: Global Warming Policy Foundation.

Ripple, W.J., Wolf, C., Newsome, T.M., Barnard, P. and Moomaw, W.R. (2020) World scientists' warning of a climate emergency. *BioScience*. 70(1): 8–12.

Robinson, K.S. (2017) *New York, 2140*. London: Orbit.

Rockström, J., Steffen, W., Noone, K. and co-authors (2009) A safe operating space for humanity. *Nature*. 461: 472–474.

Ross, A. (1991) Is global culture warming up? *Social Text*. 28: 3–30.

Rudiak-Gould, P. (2012) Promiscuous corroboration and climate change translation: A case study from the Marshall Islands. *Global Environmental Change*. 22(1): 46–54.

Rudiak-Gould, P. (2013) 'We have seen it with our own eyes': Why we disagree about climate change visibility. *Weather, Climate and Society*. 5(2): 120–132.

Rusin, N.P. and Flit, L.A. (1962) *Methods of Climate Control*. Moscow, Russia: Sovetskaya Rossiya.

Russill, C. (2015) Climate change tipping points: Origins, precursors, and debates. *WIREs Climate Change*. 6(4): 427–434.

Ryghaug, M. and Skjølsvold, T.M. (2010) The global warming of climate science: Climategate and the construction of scientific facts. *International Studies in the Philosophy of Science*. 24(3): 287–307.

Salmond, A. (2017) *Tears of Wangi: Experiments Across Worlds*. Auckland, NZ: Auckland University Press.

Sato, M. and Laing, T. (2020) Are carbon markets the best way to address climate change? YES: Markets are flexible, efficient and politically feasible. Chapter 6 (pp. 83–88) In: *Contemporary Climate Change Debates: A Student Primer*. Hulme, M. (ed.). Abingdon: Routledge.

Schneider, B. and Walsh, L. (2019) The politics of zoom: Problems with downscaling climate visualisations. *GEO: Geography and Environment*. 6(1): e00070.

Schneider-Mayerson, M., von Mossner, A.W. and Majecki, W.P. (2020) Empirical ecocriticism: Environmental texts and empirical methods. *ISLE: Interdisciplinary Studies in Literature and Environment*. 27(2): 327–336.

Scoones, I. and Stirling, A. (eds.) (2020) *The Politics of Uncertainty: Challenges of Transformation*. Abingdon: Routledge.

Scoones, I., Stirling, A., Abrol, D. and co-authors (2020) Transformations to sustainability: Combining structural, systemic and enabling approaches. *Current Opinion in Environmental Sustainability*. 42: 65–75.

Scotford, E., Peeters, M. and Vos, E. (2020) Is legal adjudication essential for enforcing ambitious climate change policies? Chapter 13 (pp. 191–206) In: *Contemporary Climate Change Debates: A Student Primer*. Hulme, M. (ed.). Abingdon: Routledge.

Scott, J.C. (1998) *Seeing Like a State: How Certain Schemes to Improve the Human Condition Have Failed*. New Haven, CT: Yale University Press.

Scranton, R. (2019) *Learning to Die in the Anthropocene: Reflections on the End of a Civilisation*. San Francisco, CA: City Lights Publishers.

Scruton, R.V. (2012) *How to Think Seriously About the Planet: The Case for an Environmental Conservatism*. Oxford: Oxford University Press.

Selby, J., Dahi, O.S., Fröhlich, C. and Hulme, M. (2017) Climate change and the Syrian civil war revisited. *Political Geography*. 60: 232–244.

Setzer, J. and Vanhala, L.C. (2019) Climate change litigation: A review of research on courts and litigants in climate governance. *WIREs Climate Change*. 10(3): e580.

Shapin, S. (1998) Placing the view from nowhere: Historical and sociological problems in the location of science. *Transactions of the Institute of British Geographers*. 23: 5–12.

Sharon, A. (2018) Populism and democracy: The challenge for deliberative democracy. *European Journal of Philosophy*. 27(2): 359–376.

Shaw, C. and Nerlich, B. (2015) Metaphor as a mechanism of global climate change governance: A study of international politics, 1992–2012. *Ecological Economics*. 109: 34–40.

Shea, M.M., Painter, J. and Osaka, S. (2020) Representations of Pacific islands and climate change in US, UK and Australian newspaper reporting. *Climatic Change*. 161: 89–108.

Shelley, M. (1818) *Frankenstein, or the Modern Prometheus*. 1818 text published by Oxford University Press, 2009.

Shelley, P.B. (1813) *Queen Mab, a Philosophical Poem, with Notes*. London.

Shelley, P.B. (1820) *Prometheus Unbound. A Lyrical Drama in Four Acts*. London: C & J Ollier.

Silko, L.M. (1992) *The Almanac of the Dead*. New York: Simon & Schuster.

Simonetti, C. (2019) Weathering climate: Telescoping change. *Journal of the Royal Anthropological Institute*. 25: 241–264.

Skelton, M., Porter, J.J., Dessai, S., Bresch, D.N. and Knutti, R. (2019) Customising global climate science for national adaptation: A case study of climate projections in UNFCCC's national communications. *Environmental Science & Policy*. 101: 16–23.

Skene, K. (2020) *AI and the Environmental Crisis: Can Technology Really Save the World?* Abingdon: Routledge.

Smith, G. and Setälä, M. (2018) Mini-publics and deliberative democracy. In: *The Oxford Handbook of Deliberative Democracy*. Bächtiger, A., Dryzek, J.S., Mansbridge, J. and Warren, M. (eds.). Oxford: Oxford University Press.

Sommer, L.K. and Klöckner, C.A. (2021) Does activist art have the capacity to raise awareness in audiences? A study on climate change art at the ArtCOP21 event in Paris. *Psychology of Aesthetics, Creativity and the Arts*. 15(1): 60–75.

Sörlin, S. (2011) The anxieties of a science diplomat: Field coproduction of climate knowledge and the rise and fall of Hans Ahlman's 'polar warming'. *Osiris*. 26(1): 66–88.

Sovacool, B.K. and Linner, B.O. (2016) *The Political Economy of Climate Change Adaptation*. London: Palgrave Macmillan.
Standke-Erdmann, M. and Viehoff, A. (2020) A world transformed by climate justice. *New Perspectives*. 28(3): 347–365.
Steffen, W., Richardson, K., Rockström, J. and co-authors (2020) The emergence and evolution of earth system science. *Nature Reviews: Earth & Environment*. 1: 54–69.
Stehr, N. (1997) Trust and climate. *Climate Research*. 8(3): 163–169.
Stenmark, L.L. (2015) Storytelling and wicked problems: Myths of the absolute and climate change. *Zygon: Journal of Religion & Science*. 50(4): 922–936.
Stephens, J.C. and Nemet, G. (2020) Should future investments in energy technology be limited exclusively to renewables? Chapter 7 (pp. 96–107) In: *Contemporary Climate Change Debates: A Student Primer*. Hulme, M. (ed.). Abingdon: Routledge.
Stern, N. (2006) *The Economics of Climate Change: The Stern Review*. Cambridge: Cambridge University Press.
Streeby, S. (2018) *Imagining the Future of Climate Change: World-Making Through Science Fiction and Activism*. Berkeley, CA: University of California Press.
Strengers, Y. and Maller, C. (2017) Adapting to 'extreme' weather: Mobile practice memories of keeping warm and cool as a climate change adaptation strategy. *Environment & Planning (A)*. 49(6): 1432–1450.
Swift, E. (2018) *Chesapeake Requiem: A Year with the Watermen of Vanishing Tangier Island*. New York: Dey Street Books and Harper Collins.
Symons, J. (2019) *Ecomodernism: Technology, Politics and the Climate Crisis*. Cambridge: Polity Press.
Thaler, R.H. and Sunstein, C.R. (2009) *Nudge: Improving Decisions About Health, Wealth and Happiness*. London: Penguin.
Thornton, T.F. and Malhi, Y. (2016) The trickster in the Anthropocene. *The Anthropocene Review*. 3(3): 201–204.
Todd, Z. (2016) An Indigenous feminist's take on the ontological turn: 'ontology' is just another word for colonialism. *Journal of Historical Sociology*. 29(1): 4–22.
Trojanow, I. (2011) *Melting Ice*. Münich: Deutscher Taschenbuch.
Tsing, A.L. (2005) *Friction: An Ethnography of Global Connection*. Princeton, NJ: Princeton University Press.
Tyszczuk, R. (2019) A brief history of scenarios. (pp. 15–24) In: *Culture and Climate Change: Scenarios*. Tyszczuk, R., Smith, J. and Butler, R. (eds.). Cambridge: Shed.
Tyszczuk, R., Smith, J. and Butler, R. (eds.) (2019) *Culture and Climate Change: Scenarios*. Cambridge: Shed.
Urgenda Foundation v The Netherlands (2015) The Hague District Court, Available at http://uitspraken.rechtspraak.nl/inziendocument?id=ECLI:NL:RBDHA:2015:7196 [Accessed 24 June 2020].
Van Den Bergh, J.C.J.M. (2017) A third option for climate policy within potential limits to growth. *Nature Climate Change*. 7(2): 107–112.
Vaughan, C. and Dessai, S. (2014) Climate services for society: Origins, institutional arrangements, and design elements for an evaluation framework. *WIREs Climate Change*. 5(5): 587–603.
Veldman, R. (2019) *The Gospel of Climate Scepticism: Why Evangelical Christians Oppose Action on Climate Change*. Berkeley: University of California Press.

Velie, A.R. (1991) *American Indian Literature: An Anthology* (2nd ed.). Norman, OK: University of Oklahoma Press.

Victor, D.G. and Kennel, C.F. (2014) Ditch the 2°C warming goal. *Nature*. 514: 30–31.

Von Neumann, J. (1955) Can we survive technology? *Fortune*. 91(6): 32–47.

Wagner, P. (2016) *Progress: A Reconstruction*. Cambridge: Polity Press.

Wallace-Wells, D. (2019) *The Uninhabitable Earth: A Story of the Future*. London: Allen Lane.

Walsh, L. (2013) *Scientists as Prophets: A Rhetorical Genealogy*. New York: Oxford University Press.

WCED (1987) *Our Common Future. The World Commission on Environment and Development*. Oxford: Oxford University Press.

Weatherhead, E., Gearheard, S. and Barry, R.G. (2010) Changes in weather persistence: Insight from Inuit knowledge. *Global Environmental Change*. 20(3): 523–528.

Webber, S. (2015) Mobile adaptation and sticky experiments: Circulating best practices and lessons learned in climate change adaptation. *Geographical Research*. 53(1): 26–38.

Webber, S. and Donner, S.D. (2017) Climate service warnings: Cautions about commercializing climate science for adaptation in the developing world. *WIREs Climate Change*. 8(1): e424.

White Jr., L. (1967) The historical roots of our ecologic crisis. *Science*. 155: 1203–1207.

White, S. (2015) "Shewing the difference between their conjuration, and our invocation on the name of god for Rayne": Weather, prayer, and magic in early American encounters. *The William and Mary Quarterly*. 72(1): 33–56.

Whyte, K. (2020) Too late for indigenous climate justice: Ecological and relational tipping points. *WIREs Climate Change*. 11(1): e603.

Wilkinson, K.K. (2012) *Between God and Green: How Evangelicals are Cultivating a Middle Ground on Climate Change*. New York: Oxford University Press.

Wilson, E.K. (2012) *After Secularism: Rethinking Religion in Global Politics*. Basingstoke: Palgrave Macmillan.

Wolf, J. and Moser, S.C. (2011) Individual understandings, perceptions and engagement with climate change: Insights from in-depth studies across the world. *WIREs Climate Change*. 2(4): 547–569.

INDEX

Note: The index covers the main text but excludes Class Questions, the Bibliography and material only to be found in each chapter's Further Reading suggestions. Page references in *italic* and **bold** type are used to distinguish relevant material in *Figures* and **Tables** respectively: the suffix 'n' indicates material in chapter endnotes.

For Product Safety Concerns and Information please contact our EU
representative GPSR@taylorandfrancis.com
Taylor & Francis Verlag GmbH, Kaufingerstraße 24, 80331 München, Germany

www.ingramcontent.com/pod-product-compliance
Ingram Content Group UK Ltd.
Pitfield, Milton Keynes, MK11 3LW, UK
UKHW020131200926
475702UK00005B/689
* 9 7 8 0 3 6 7 4 2 2 0 2 8 *